Introduction to COMMUNICATIONS TECHNOLOGIES

A Guide for Non-Engineers

Third Edition

OTHER COMMUNICATIONS BOOKS FROM AUERBACH

Introduction to COMMUNICATIONS TECHNOLOGIES

A Guide for Non-Engineers

Third Edition

Stephan S. Jones
Ronald J. Kovac
Frank M. Groom

CRC Press is an imprint of the
Taylor & Francis Group, an **informa** business
AN AUERBACH BOOK

CRC Press
Taylor & Francis Group
6000 Broken Sound Parkway NW, Suite 300
Boca Raton, FL 33487-2742

Printed on acid-free paper
Version Date: 20150629

International Standard Book Number-13: 978-1-4987-0293-5 (Hardback)

Visit the Taylor & Francis Web site at
http://www.taylorandfrancis.com

and the CRC Press Web site at
http://www.crcpress.com

Contents

Preface

Our original effort in writing this book was to create a starting point for those in the business community who did not have a high level of technical expertise but needed to have some understanding of the technical functions of their information and communication technologies (ICT) in a corporate environment. As was true with the first edition of this book, if you are already an engineer, find some other form of pleasure reading—this text is not designed for you!

The third edition of *Introduction to Communications Technologies: A Guide for Non-Engineers* has been updated to attempt to keep pace with the ever-changing industry associated with ICT. The basic fundamentals of communicating information have not changed: system models and electricity are still the building blocks of how we send information around the world. However, with the advancement of faster processors within our communicating devices, we have seen a rapid change in how information is modulated, multiplexed, managed, and moved!

We have also tried to include in this edition some forward perspectives on where the networks we currently use will migrate in the next few years. With streaming video, Internet Protocol–defined voice communications, advanced wireless data networking, and a convergence of our various communication methods (i.e., voice, data, and media) to be delivered on a single platform, various sections address this changing landscape. As an example, the chapter on Multiprotocol Label Switching (MPLS) gives the reader a valuable understanding of what should be expected from vendor services for external networking offerings in the near term, of how the core of the network is changing, and of how traffic engineering is impacted by MPLS-defined virtual private networks (VPNs).

As with our previous editions, we have made a concerted effort to avoid trying to confuse readers with too many equations or having them perform calculus gymnastics. These formulas and functions are critical in creating the granular components and operations of each technology; however, understanding the application and its purpose to the business environment does not require a scientific calculator. Reading this text should provide a knowledgeable starting point for further exploration into any areas impacting the reader's communication challenges.

We use this text in our core curriculum to help students without technical undergraduate degrees entering our graduate professional program in information and communication sciences to get a good, broad understanding of what they should expect from their educational experiences. The students have consistently requested that the book be made a "mandatory" summer read, prior to their showing up for classes. We hope that the third edition continues to provide the reader with the knowledge necessary to work within this fast-paced and ever-changing industry.

Student Acknowledgments

Our thanks to the following individuals for their assistance and contributions to this text:

William Brunson
Jessica Chaffin
Jonathan Chamboneth
Matt Cresswell
Larissa Denton
Charles Geyer
Rebekah Hobbs
Maria Lee
Breanna Parker
Kalyn Sprague
Steven Williams

Authors

Stephan S. Jones, PhD, joined the Center for Information and Communication Sciences faculty in August of 1998. He came to Ball State University (BSU), Muncie, Indiana, after completing his doctoral studies at Bowling Green State University, Bowling Green, Ohio, where he served as the dean of continuing education, developing a distance learning program for the College of Technology's undergraduate technology education program. Dr. Jones was instrumental in bringing the new program on board because of his technical background and extensive research in the distance learning field.

Prior to coming to higher education, Dr. Jones spent over 16 years in the communication technology industry. He owned his own interconnect, providing high-end commercial voice and data networks to a broad range of end users. Dr. Jones provided all the engineering and technical support for his organization, which grew to over 20 employees and became one of the top 100 interconnects in the country. Having sold his portion of the organization in December of 1994, Dr. Jones worked for Panasonic Communications and Systems Company as a district sales manager providing application engineering and product support to distributors in a five-state area.

Since coming to BSU, Dr. Jones has been engaged in development efforts to provide up-to-date equipment for the graduate students' production labs, the Network Integration Center, and the Wireless Institute's lab. He has also partnered with local nonprofits to provide community support for their information technology needs through graduate student service learning projects. He has appeared as a presenter at regional development gatherings and has provided research agendas to facilitate the application of broadband technologies for underserved constituencies. In his new role as the director of the graduate program, he is charged with making the success of the past 20 years viable for the next generation of information and communication technology students.

Frank M. Groom, PhD, is a professor in the Graduate Center for Information and Communication Science at Ball State University, Muncie, Indiana. His research is concentrated in the areas of high-bandwidth networking, distributed systems, and the storage of multimedia objects. Dr. Groom is the author of seven books, most recently having finished *The Basics of Voice over IP Networking* and *The Basics of 802.11 Wireless LANs*. Among his best known books are *The Future of ATM* and *The ATM Handbook*. Dr. Groom earned his PhD from the University of Wisconsin–Milwaukee in information systems. He is the former senior director of information systems for Ameritech.

Ronald J. Kovac, PhD, is a full professor in the Center for Information and Communication Sciences at Ball State University, Muncie, Indiana. The center prepares graduate students in the field of telecommunications. Previous to this position, Dr. Kovac was the telecommunication manager for the state of New York and a CIO for a large computing center located on the east coast. Dr. Kovac's previous studies included electrical engineering, photography, and education. Dr. Kovac has published two books and over 50 articles and has completed numerous international consulting projects in both the education and telecommunications field. Additionally, he speaks worldwide on issues related to telecommunications and holds numerous certifications, including the Cisco Certified Network Associate (CCNA), the Cisco Certified Academy Instructor (CCAI), and the almost-complete Cisco Certified Network Professional (CCNP).

Chapter 1

Systems and Models of Communications Technologies: Shannon–Weaver, von Neumann, and the Open System Interconnection Model

The language, acronyms, and terminology in communications technology (CT) are based on fundamental ideas that can be considered systems. Understanding these fundamental systems allows knowledge of various information technologies (IT) to be added, brick by brick, to build an understanding of current and developing technologies. One of the primary systems in CT is signaling. Signaling occurs in traditional voice telephone connections, local area networking, and wireless communications ranging from cellular to satellite systems. Each of these areas of signaling usually becomes the domain of an engineer who is focused specifically on this area of expertise. Preparing someone to understand all the intricacies of each of these disciplines is a huge undertaking. However, understanding the basic underlying principles of the signaling process can give the nonengineering professional the ability to converse on the topic intelligently.

The capacity to communicate is determined by a number of factors that influence the quality of the signal. How will the information be sent or transmitted? Will the person or device at the distant end of the communication signal be able to receive the transmitted signal? Will the person or device at the receiving end need to decipher or decode the signal in any way? What would happen if noise were introduced into the transmitted signal? Would it cause the transmission signal to be corrupted in any way?

Exhibit 1.1 shows how communications can be understood in terms of a model (or systems format). The source (transmitting side of communications) signaling systems need to encode the information to be transmitted so that it fits onto the medium or channel that is used to convey the information. The medium could be air, copper wires, or even fiber optics. When we speak to another human, we first determine whether or not the person understands the language we are speaking. As the source, we encode our information in the language and put it on the medium (air) to be delivered to the other person. What if the transmitting person has French as his or her primary language and the receiving person does not? The receiving person will need to decode the French language into one that he or she is able to understand. The return signal from the receiving person may ask the transmitting person, "Is this what you are trying to tell me? If it is, then we can discuss this topic further." This type of transmit/receive/confirm format is the basis of human communication and is also employed in data networking transmissions.

Claude Shannon developed a mathematical equation that defines the theoretical limit of the capacity of information transmitted in this model. He theorized (and later proved) that the amount of information being transmitted was based on a number of factors, including noise, frequency, transmission, and the strength of the signals. The formula is

$$C = B\log_2(1 + S/N)$$

Do not be concerned about solving the math problem! It is presented to show the relationship between the signal (S) and noise (N) in a given transmission based on the bandwidth (B) or frequency at which the signal is being transmitted. The capacity (C) of information being transmitted is determined by all of these factors. Later, when we consider more complex topics, we will be able to use this formula to help us understand how much information can be transmitted over a wireless connection for our local area network or how our cell phone can be used to connect to

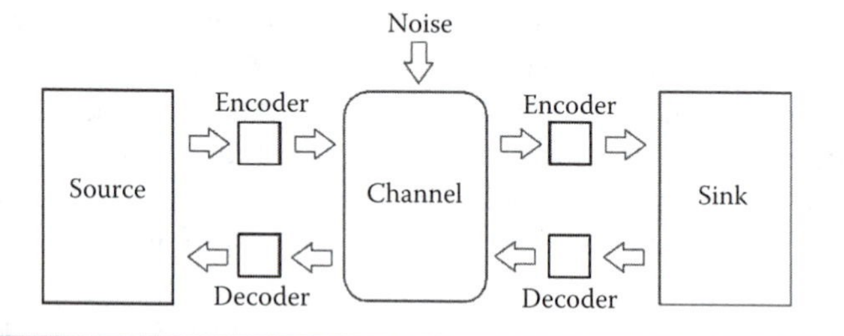

Exhibit 1.1 Shannon–Weaver model of communication.

the Internet. This formula tells us that as the noise in a transmission increases, the capacity to send information decreases. As the frequency (or bandwidth) at which we transmit increases, the capacity for information transfer increases. We will discuss the topics of frequency and bandwidth in Chapter 3.

Computing Model

Any device on a communication network can be considered a *node*. A node can be a computer, telephone, router, server, tandem switch, or any number of devices that receive and transmit information on their respective networks. If every device connected to a network used a different format for collecting, storing, modifying, or transmitting information, the design of networks and the components connected to them would be extremely difficult to accomplish. John von Neumann, a mathematician whose theories led to the development of the first electronic digital computer, proposed a model based on work originally presented by Alan Turing (a mathematician who is famous for leading the group responsible for deciphering the Enigma code during World War II). Von Neumann's idea is known as the *stored program model*. Exhibit 1.2 is a simple block diagram of von Neumann's model.

The general structure of this model is based on four primary components, and a fifth component is necessary for interblock signaling. This model can be extrapolated to define how the nodes connected to any communication network operate:

1. Main memory: stores data and instructions
2. Arithmetic Logic Unit (ALU): performs computational functions on binary data
3. Control unit: interprets the instructions in memory and causes them to be executed
4. Basic Input/Output System (BIOS): devices operated by the control unit

The fifth component of this model is the bus structure. The information that needs to be exchanged between these blocks relies on an interconnecting medium referred to as a *bus*. Bus structures can be found in everything from digital wristwatches to the most sophisticated high-speed computer device. We will look at this model and

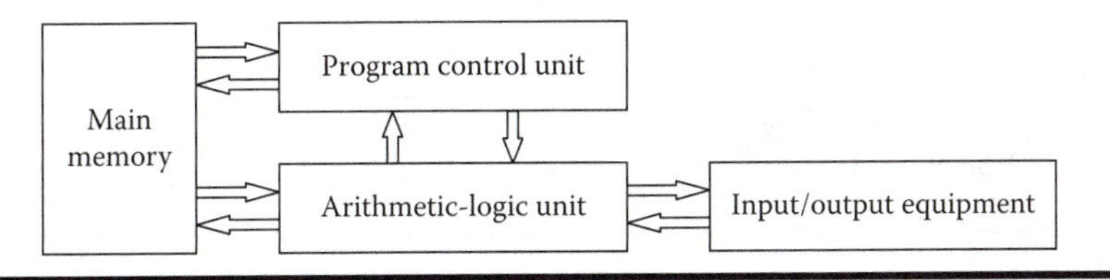

Exhibit 1.2 Components of the stored program model.

how it relates to a computer and, using the same model, define how it works with telephone systems, data networking equipment, and other nodal devices.

The main memory on the computer is its working area. The program files that are needed to run specific applications are stored in this work area. An analogy is with a large legal pad of paper. The program running on a computer occupies the top page of a writing tablet: this area of memory is also called the *random access memory* (RAM) of a computer. If it needs more information to process a request, the system will look into the pages available on the tablet to see if the data are there. To retrieve more information for a program or to pull a completely new program into the work area, the system needs to retrieve more pages of information (data). Another typical memory storage area is the hard drive, which, according to our model, is viewed as an I/O device.

The process of responding to requests for data is based on a set of instructions written in binary language. Most of us have never seen the machine language that runs our digital devices, but it is the mother tongue of the information age. Requests for printer or modem access (instructions for information retrieval) and other functions are acted upon by the central processing unit (CPU). The CPU relies on the ALU to crunch numbers and feed the information back so that it can control the flow of data.

Exhibit 1.3 represents a more detailed view of the stored program model. The memory portion of the diagrams is now populated with the various pages of information necessary to operate the computer: BIOS, operating system, and application programs. The bus structure represented by arrows in Exhibit 1.2 is shown in

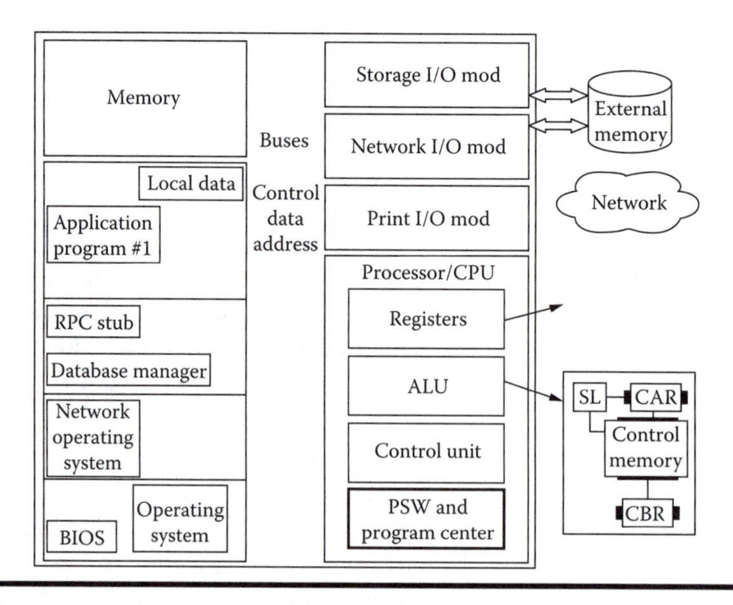

Exhibit 1.3 von Neumann's architecture in a computing device. (Courtesy of Frank M. Groom, Movement and Storage of Information, class notes.)

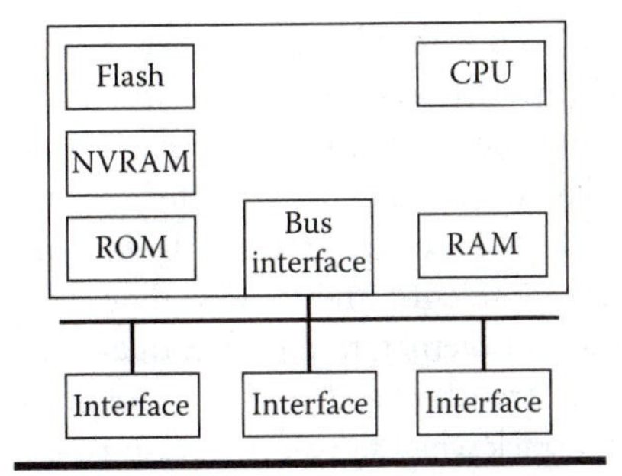

Exhibit 1.4 Simple router architecture block diagram.

greater detail in Exhibit 1.3, which shows the connection between the memory, CPU, and ALU. The I/O modules are connected to external memory (hard drives), networks, or printers. This greater level of detail depicted in Exhibit 1.3 is also applicable to other network equipment that requires processing and memory management to work.

This model can be used to define how a router, digital personal automatic branch exchange (PABX), and data networking switches work. (These topics are defined in subsequent chapters.) A router is a device used in data networks to forward information from one network to another. They are the workhorses of the Internet, the World Wide Web (the web), and corporate wide area networks (WANs). With this level of responsibility, the equipment may seem more complex and sophisticated than it really is. A router is required to analyze packetized information received from its I/O module connection to decide where the information packet is sent or routed next. This requires an operating system that keeps the organization of the system flowing correctly. A CPU to control requests in conjunction with the ALU is necessary to analyze the binary data. Connecting all of these parts together is a bus structure that is able to send and receive millions of requests every second!

Exhibit 1.4 is another example of how the von Neumann architecture works on a different type of computing device. The router has all the components previously defined for a computer, and it works in a similar manner. The operation system and application programs reside in the flash and other nonvolatile memory. Just as in the computer, RAM is the working area memory of the router. A router does not need the same memory capacity as a computer, but it requires the same binary process, or 1s and 0s, to analyze data.

By understanding the von Neumann architecture, you can comprehend how any number of devices can operate on different networks. The foundational work for this model was presented in 1946; 61 years later, it continues to be used to define the operation of digital devices.

Open System Interconnection Model

The ability to interchange vendors' products in the network without issues of incompatibility has long been a vital concern of those who work in the industry. Most vendors strive to enable various components to talk intelligently to one another reliably and consistently. The International Standards Organization (ISO), whose members are representatives from countries around the world, has developed the Open System Interconnection (OSI) model. This model consists of seven layers. The OSI model, although not used as a *de facto* standard by all companies developing networking equipment today, is used as a guideline to simplify the development of interoperability among vendors. The most consistent adherence to the OSI model is found at its lower levels (1 through 4). Adherence is less in the higher layers of the stack. Some of this information may not make a great deal of sense to you initially but will become clearer as we progress through different technologies and see how they communicate with the OSI model. Technologies such as Ethernet, frame relay, Integrated Services Digital Network (ISDN), Asynchronous Transfer Mode (ATM), and other confusing acronyms are all compared to the OSI stack when signaling, transmission, encryption, and other functions are discussed. Understanding the basic principles of the model is important at this point in your reading.

Exhibit 1.5 shows each layer's role in the model and a brief definition of its function. A more in-depth examination of each layer is required to fully understand the purpose of, and interrelationships among, the layers. Layer 1 is the physical layer. It is used to describe how data that will be sent out on the network will be transmitted. It defines electrical, optical, or frequency modulation, depending on the type of medium (e.g., copper, fiber, wireless) the network employs. Connection standards such as EIA-232 are defined in this layer. This physical layer should not be confused with the actual cable used for transmitting the data; it is the layer within the OSI model that defines how that connection will be made. Modem connections to the coaxial cable are defined within this layer.

The data-link layer is the second layer of the OSI model. This layer is concerned with physical (as opposed to logical) addressing. At this layer, a network device (e.g., a computer) has an address assigned to its network interface card (NIC) that uniquely identifies a device. Layer 2 is also concerned with network access (e.g., the NIC), framing of upper-layer data for use at the physical layer, and error detection. Layers 1 and 2 are the most universally accepted protocol layers of the OSI model.

	Layer	Function
7	Application	User interface: what we are doing
6	Presentation	Form, syntax, language, encryption, compression
5	Session	Resource control, handshaking, bind/unbind
4	Transport	End-to-end addressing, segmentation, muxing
3	Network	Routing, error recovery, packetization, IP addressing
2	Data-link	Point to point, error-free physical layer, framing
1	Physical	Mechanical, electrical, logical to physical, bits

Exhibit 1.5 Seven layers of the OSI Reference Model.

The third layer, the network layer, has become an integral cog in the movement of data across worldwide internetworking. The logical addressing of a device is located at this layer. Routing of information on the Internet, the web, and other WANs is based on the logical address found at the network layer. The most widely used and accepted layer 3 addressing scheme is the Internet Protocol (IP). (Path or route selection is based on the IP address of a device.) The network layer is also responsible for formatting data from higher layers into packets that are passed down to the data-link layer for organization into frames to be delivered to the physical layer.

At this point, it is important to note that the network layer does not interact with the physical layer. Layer 3 does not know what signaling format layer 1 will use to provide data to whatever medium is being used to deliver the information. Each layer of the OSI model only interacts with the layer directly adjacent to it in the stack. Encapsulation, the process of putting received information into another format for delivery, occurs at each layer of the OSI stack. This allows for flexible architectural arrangements and the ability to change layer 1 and layer 2 programming to accommodate the connectivity of disparate systems. Think of a letter addressed in French to someone on a university campus. Because the handlers of the mail may not understand French, the letter is put inside another envelope addressed in English so that the university's postal network can easily identify it. The English-labeled envelope is the encapsulation of the French data. It is this flexibility that allows an Apple computer to be networked with Windows-based machines.

The transport layer, the fourth layer of the model, is primarily responsible for end-to-end integrity of communications between two nodes on a network. The transport layer establishes, manages, and terminates the exchange between two devices. End-to-end error correction and the flow control of data transmission are the responsibility of the transport layer. Layer 4 is responsible for the segmentation of upper-layer data being sent down for external delivery. The segmented data are passed down to the network layer to be placed in packets. Transport Control Protocol, or TCP, resides at this layer. TCP is responsible for the integrity of the data delivered with IP. The two protocols—TCP and IP—are the primary carriers of data today. Other delivery protocols live at this layer and will be discussed when TCP/IP is defined in a subsequent chapter.

The next three layers of the OSI model are generally considered the realm of software design, in contrast to the first four layers, which relate to the hardware configuration of a device. The upper layers have not been developed universally and may not even be defined for some devices or connections, unlike the lower layers.

The session layer, the fifth layer, is responsible for establishing, managing, and terminating communications between two devices. It is also the area in which full-duplex or half-duplex communication is defined. The session layer determines if one device can be interrupted by another device while communicating. If a session connection is lost during data transfer, the session layer is responsible for helping recover the communication.

Layer 6, the presentation layer, is the translator and interpreter for data being sent from the upper application layer and for the data moving up the stack to be acted on by the application layer. The presentation layer is also responsible for encryption, which is defined as the conversion of data into a secure format that can be read only by predefined recipients. Compression formats are also found at this layer. In the recent past, the sixth layer has been the least defined layer of the model. However, with the advent of new video and voice technologies offered over the internetworks, this layer is becoming a critical component for communication across the WAN.

The last layer, layer 7, is the application layer, which is the closest to the user. It is the interface that lies between the network and the application being used on a device. E-mail protocols, the Hypertext Transfer Protocol (HTTP) web language, and other interface applications reside at this layer. Error recovery from data being assembled after transmission also can be the responsibility of the application layer. It provides for a final integrity check of the received data transmission.

The OSI model has been put to use in developing numerous advanced networking technologies and continues to provide the technical mapping for integration between disparate networks, vendors, and software. This model simplifies the understanding of how complex sets of protocols interact to deliver information across the Internet. One of the most widely used protocols based on this model is the TCP/IP suite. Drawing an analogy between the delivery of information over the Internet and the writing, addressing, and eventual delivery of a letter and comparing the process with the OSI model are helpful in understanding how both TCP/IP and the OSI model work.

When writing a letter, a medium for delivery has to be selected. A blank sheet of paper is used to record information in a written language that is understandable to the reader. When the letter is completed, it is placed in an envelope and addressed; the postal service routes the letter to its destination. Drawing an analogy with the TCP/IP protocol, data are created in the upper layers of the OSI model (blank sheet of paper) and prepared for sending. They are encapsulated (the envelope) as they pass through layers 4 and 3. At layer 3, the data are given a logical address that is understood by the delivery system (postal service). When the data are pushed out to the network, the address is used to route (deliver) the information to its destination. The addressing scheme associated with TCP/IP is a simple format similar to a home address but sophisticated enough to provide millions of addresses understood all over the world. A more detailed look at this addressing scheme is explored in Chapter 9 on WANs.

Summary

The key to understanding complex systems lies in learning the fundamental platforms on which they are built. Using these models as a template to figure out complex technology interactions is a functional form of problem solving for the novice as well as the experienced technologist. Shannon's communication model defines how much information can be processed, given the parameters of the environment in which the data are to be transferred. This model can be employed in wire line as well as wireless delivery schemes. It can also be used to define how humans communicate in various settings.

von Neumann's stored program model gives us an idea of how the majority of computing devices have been configured since the mid-1940s. The model can be used to understand how other nodal components of networks, such as routers, switches, and newer integrating devices that combine voice, data, and multimedia into a single network, function. The five basic components of the model have different capacities and function in all these devices.

The OSI model is the most widely used generic protocol stack for communications over various networks. The OSI model was developed by an international group that wanted to create interoperability across disparate networks, manufacturers, and vendors. The TCP/IP suite of protocols is modeled from this stack. It is the primary delivery mechanism for information over the Internet. Its addressing scheme is understood worldwide and continues to evolve as newer technologies continue to emerge.

Review Questions

Multiple Choice

1. According to the capacitance theorem, as the noise in a transmission increases:
   a. The capacity to send information increases
   b. The capacity to send information decreases
   c. The capacity to send information is not affected
   d. None of the above
2. All of the following are components of the von Neumann stored program model, except:
   a. Main memory
   b. Program control unit
   c. ALU
   d. OSI
3. The OSI model consists of how many layers?
   a. 4
   b. 5
   c. 6
   d. 7
4. A node is:
   a. Any device on a communication network
   b. A computer
   c. A printer or a computer
   d. A printer
5. According to the von Neumann model, a hard drive is:
   a. A part of the main memory
   b. A part of the ALU
   c. An input/output device
   d. A part of the bus structure

6. Encapsulation is:
   a. Something that occurs at layer 5 only
   b. The process of putting information into another format for delivery
   c. Something that occurs at each layer of the OSI model
   d. Both b and c
7. Internet Protocol (IP) resides at what level of the OSI model?
   a. The transport layer
   b. The network layer
   c. Layer 2
   d. The data-link layer
8. All of the following are components of the Shannon–Weaver model of communication, except:
   a. Encode/decode
   b. Message
   c. Capacity
   d. Noise
9. The OSI model was created by:
   a. Shannon
   b. von Neumann
   c. The International Standards Organization (ISO)
   d. The Institute of Electrical and Electronic Engineers (IEEE)
10. In the OSI model, each layer:
    a. Interacts with all other layers
    b. Only interacts with the layers directly above or below it
    c. Does not interact with any other layers
    d. Only interacts with layers that are higher than it

Matching Questions

Match the following terms with the best answer:

1. $C = B\log_2(1 + S/N)$	a. A component of the Shannon–Weaver model
2. Noise	b. Model containing seven layers
3. Data-link layer	c. TCP
4. ALU	d. Frames
5. Transport layer	e. Developed the stored program model
6. Packetization	f. Occurs at all levels of the OSI model
7. von Neumann	g. Capacitance theorem
8. OSI	h. Occurs in layer 3 of the OSI model
9. I/O	i. Performs computational functions on binary data
10. Encapsulation	j. Operated by the control unit in the von Neumann model

Short Essay Questions

1. Explain the communication process in terms of the Shannon–Weaver model.
2. Why is it necessary to have a standard model with which to discuss networks or computers?
3. List and define each component of the von Neumann stored program model.
4. List and explain what occurs at each layer of the OSI model.
5. Explain the impact that the von Neumann model has had in terms of digital devices.
6. Using examples, explain each step of the communication process in terms of the Shannon–Weaver model of communication.
7. Define the fifth component of the stored program model, and explain how it is used.
8. Explain, in words, the capacitance theorem.
9. Think of a situation where you think that knowledge of the systems presented in this chapter would be helpful in your career.
10. Explain why it is important for a nonengineering professional to understand the material presented in this chapter.

Chapter 2

Basic Concepts of Electricity

An appreciation and understanding of communications technologies is attainable without ever delving into hard-core engineering topics of physics and calculus. An understanding of the applied nature of the components of electricity, however, will give nonengineering professionals a greater understanding of the topics covered in this book. Explaining the associated terminology with examples of their use in networking environments will help make the connection between theoretical and practical science.

This chapter will cover the basic concepts of electrical systems in telecommunications, the functions associated with electrical components (e.g., resistor, capacitor), and their effect on signaling (e.g., resistance, capacitance).

Common Units of Technical Measurement

Conversations between technically oriented people invariably employ terminology that is foreign to those who do not understand the nomenclature of various measures used in communications technology. When the throughput of a circuit is described as 1.5 Mbps, it is necessary to know that M denotes the prefix *mega*, which stands for the value 1,000,000, or 10^6. Understanding these units is important when comparing different service offerings, evaluating mean time between failures of various components, or reading the calculated aggregation of network throughput necessary for purchasing a circuit for a corporate office connection. Exhibit 2.1 gives a brief description of the prefixes and their associated values.

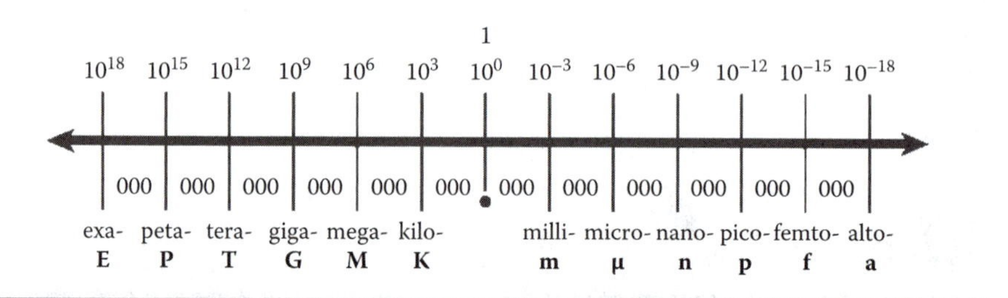

Exhibit 2.1 Scientific notation chart.

Prefix	*Frequency*	*Data*	*Voltage*	*Resistance*	*Power*
Tera	Terahertz (THz)	Terabit (Tb)			
Giga	Gigahertz (GHz)	Gigabit (Gb)			
Mega	Megahertz (MHz)	Megabit (Mb)	Megavolt	Megohm	Megawatt
Kilo	Kilohertz (kHz)	Kilobit (kb)	Kilovolt	Kiloohm	Kilowatt
Milli			Millivolt	Milliohm	Milliwatt
Micro			Microvolt	Microhm	Microwatt

Exhibit 2.2 Common uses of scientific notation.

Exhibit 2.1 also provides a reference for the various names and values associated with the power of 10 (10^n). Each place represents three 0s to be added or subtracted from the number being discussed, depending on the direction away from one the value is moving. The most common uses of the prefixes are listed in Exhibit 2.2.

These terms will be used throughout this book without further explanations of their meaning. The exhibits will make excellent reference tools for further reading in communications technologies.

Signals

The delivery of information across voice and data networks requires a signal from a sender to a receiver to be sent somewhere along the communication channel to either establish the circuit in which the information is to travel or verify that information has been sent and received. Exhibit 2.3 shows a block diagram of a voice circuit. The signaling process is necessary to establish a connection between the sending party and the receiving party (remember Shannon's law?). The signals are generated in traditional wire-line (as opposed to wireless) circuits by signals or frequencies being sent from the dial pad on the telephone.

The signals are propagated over the copper wire infrastructure to the central office (CO; the telephone company switching gear closest to your premises). From there, the analog signal is converted to a digital transmission and sent out over fiber-optic connections to the various tandem switches, toll offices, Internet service

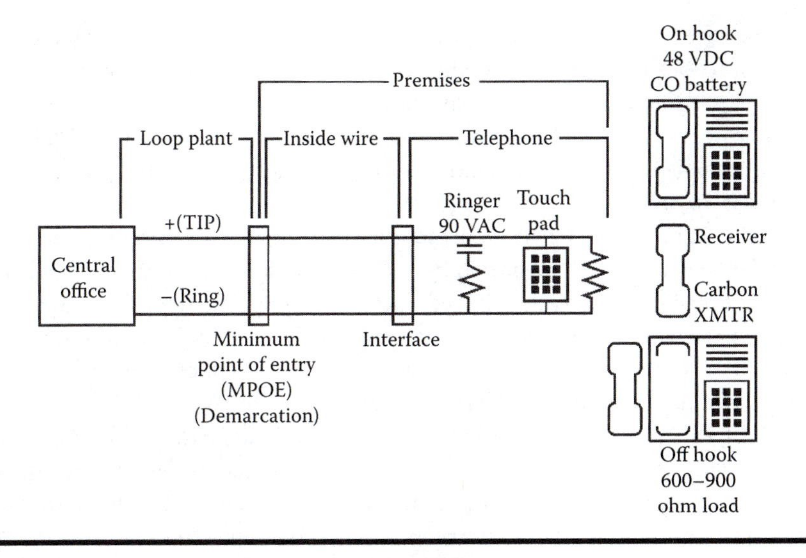

Exhibit 2.3 Components of wireline circuit.

providers (ISPs), and other entities that use the existing infrastructure for access. Over the course of this book, we will progress through topics such as the basic signaling system, the creation of the electrical portion, the traditional analog-to-digital conversion scheme, the layout of the telephone network and how it is connected, and finally, the steps involved in transmitting through the network.

A signal is typically defined by a waveform, a picture produced by an oscilloscope that shows the highs and lows of the voltage or current associated with the signal. The waveforms vary in intensity, shape, duration, and complexity. Exhibit 2.4 gives an example of a simple waveform that is graphed over a period of time in reference to its voltage output. Note that the height, or peak, of the signal is gradually diminishing, its intensity attenuating (or reducing) over time. The variation and repetition of a waveform over a specific period of time is referred to as the

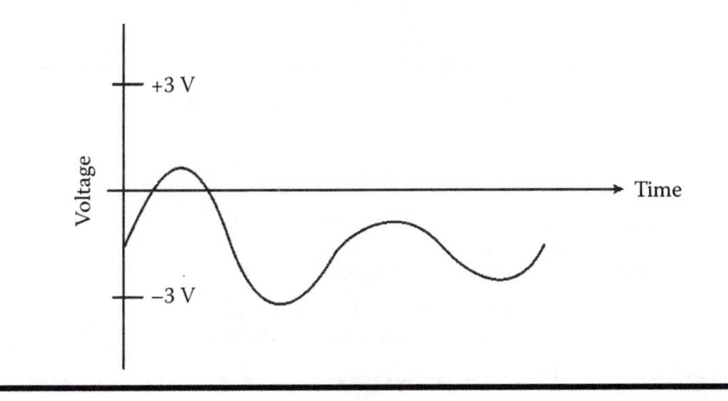

Exhibit 2.4 Simple waveform.

waveform's *cycle*. The cycle is usually measured by how many times per second the waveform replicates itself on the graph. This measure is called cycles per second or, in more current terms, hertz. If a waveform completes one cycle every second, it is said to be operating at 1 Hz; if a waveform is replicating itself at a thousand times every second, it is said to be operating at 1 kHz. Electricity in the United States operates at 60 Hz.

Exhibit 2.5 shows a signal replicating itself twice over a 1 s time period. It is said to be operating at 2 Hz, or two cycles per second. The frequency of a signal is defined as the number of cycles divided by the time in which they occur. The period is the time it takes a waveform to complete one full cycle. These waveforms are characteristic of various signals in the public switched telephone network (PSTN), wireless communications, and data networks.

Taking this principle one step further, we can discuss bandwidth. Bandwidth is the range of frequencies that a communication channel is designed for so that specific signaling functions can be accomplished. It is incorrect to refer to throughput (bits of information usually measured by number per second, e.g., kilobits per second) and bandwidth interchangeably; however, this practice is common in the industry today. Exhibit 2.6 shows a visual definition of bandwidth. The values discovered for the areas defined by subtracting the lower frequency from the higher frequency give the bandwidth for a particular band; in this case, the band is associated with bandwidth A and bandwidth C.

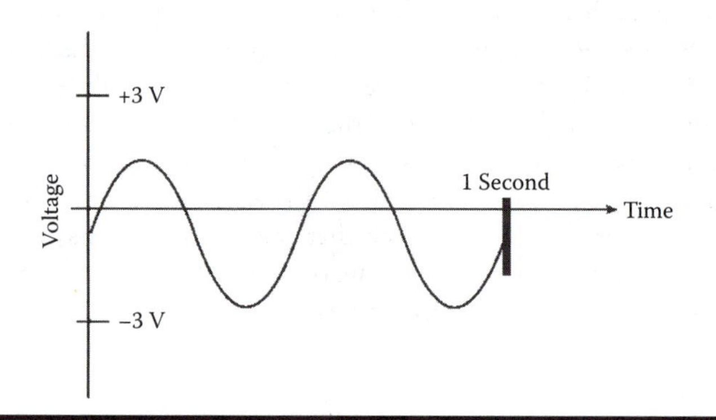

Exhibit 2.5 Repeating waveform.

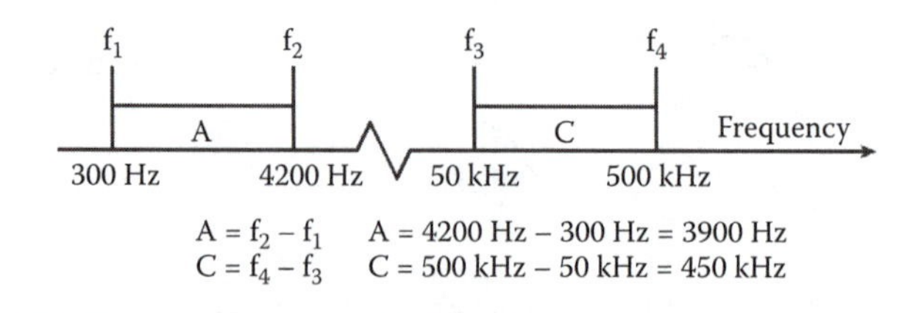

Exhibit 2.6 Bandwidth representation.

Current

One of the most feared and avoided courses in high school and college is physics. Physics is the scientific discipline that explains how things work in the universe. It is not necessary for nonengineers to go into formulas and calculus to learn how to apply the principles of physics to communications technologies. A simplistic definition of the building blocks of matter and how they relate to one another is important in our discussion of how information is transmitted from a source to a receiver.

All matter in the universe is made up of microscopic components called molecules. These molecules are the smallest definable piece of a material. Molecules can be broken down into even smaller building blocks called atoms. Atoms are made up of even smaller structures that hold different types of electrical charges: protons (positively charged particles), neutrons (no electrical charge), and electrons (negatively charged particles). Protons and neutrons make up the center of the atom, its nucleus. Orbiting around the nucleus are the electrons, usually in equal numbers in relation to the protons in the nucleus. A basic principle of physics states that the protons and neutrons are bound together by a force that is directly proportional to the size of the particle and its distance from other particles of similar and opposing electrical charges. At the outer edges of the atom, the forces that hold the electrons in orbit around the nucleus are not as strong as those closer to the center of the nucleus; these loosely held electrons are referred to as free electrons because they can be moved out of their orbit. It is the movement of free electrons that causes electric current. Exhibit 2.7 is a block diagram of what an atom with three protons and three electrons looks like under a high-powered electron microscope. Note that the outer orbiting electron is most likely to become a free electron.

When atoms lose and gain electrons, an electrical charge is created that can be harnessed to make all the various components of communications technologies function. Some materials can be made to move electrons between atoms at a more excited rate than other materials. A conductor is a material, such as copper, gold, or silver, that has a very high excitability and can carry electrical current freely. An insulator is a material that does not allow its electrons to break away from their nucleus easily. Plastic, rubber, and glass are considered good insulators.

The rapid movement of outer orbiting electrons in the atom of a conductor causes current to flow. One atom's electron jumping into the orbit of another atom, followed by another jump, and then another in sequence, is how electrons create current flow in a conductor. A by-product of this rapid movement of electrons is heat generation. If a device or conductor is not capable of handling high current, the cable will feel warm to the touch (and sometimes burn through). Heat dissipation is an important aspect

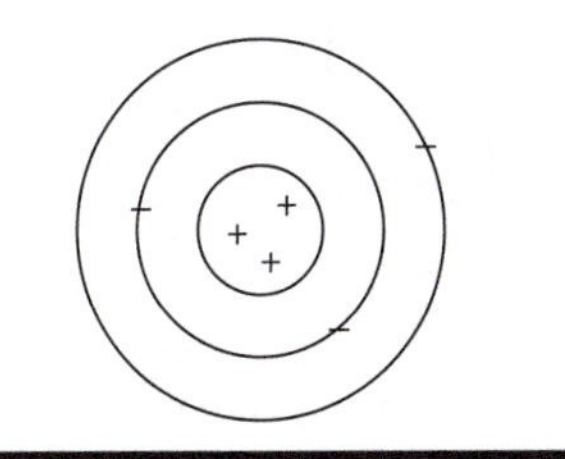

Exhibit 2.7 Simple atom structure.

of designing electronic circuitry because of the current flow through the various parts in the device.

Current flow is the movement of the free electrons in the same direction in a conductor. The unit of measure used to describe current flow is the ampere, or amp. The letter I is used to represent current in electronic schematic drawings and equations.

Alternating current (AC) is the type of electrical power available at wall outlets. It is one of the two types of current flow; the other is direct current (DC). The charge of the current passing through the conductor changes from a positive to a negative value over a period of time. In the United States, the value alternates 60 times per second; in Europe, 50 cycles per second. At this rate, a light bulb switches on and off 120 times per second—too fast for the human eye to detect. Exhibit 2.8 shows the rate of change of an AC signal. AC can be thought of as changing direction as it goes from its positive to its negative position. This process is continuous until the power source is eliminated.

DC is used primarily in electronic components in computers, cellular phones, PSTN, and data networks. It is capable of doing more work than AC; however, it does not travel as well, meaning it cannot be distributed over long distances as efficiently as AC. DC can be considered a continuous flow of current, in contrast to the periodic variation of AC. Exhibit 2.9 compares DC current to AC current.

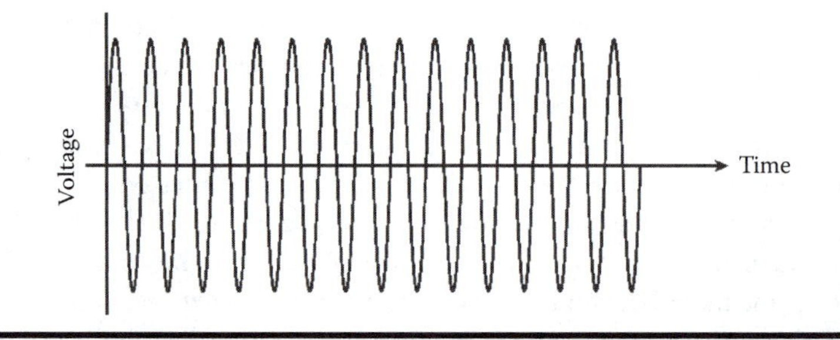

Exhibit 2.8 Alternating current over a 0.25 s time period.

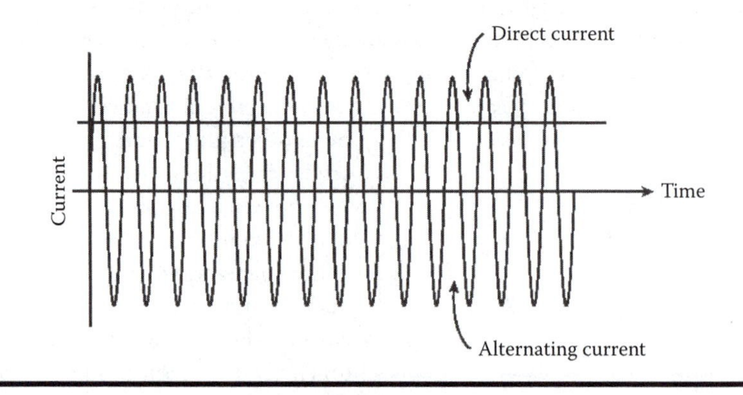

Exhibit 2.9 Direct current versus alternating current.

Resistance

Resistance is the force that opposes the flow of electrons in a material. Resistance is represented by the letter R in mathematical equations and by the symbol W (ohm) in graphical drawings or on components that need to have their resistive values noted. Every material has some form of resistance to electron flow, some more than others. These values of resistance are directly related to the previous discussion about conductors and insulators. Different values of resistance are used to vary the current flow to various components in a device. By manipulating a single source of current to service a wide variety of functions using inexpensive resistors (and other electronic devices), the cost of a device is greatly reduced. The cost of a device increases proportionately as power and current requirements become more complex.

Voltage

The ability to perform work, or the energy potential of the electrical charges described previously, is measured by the unit *volt*, which is represented by the letters V or E in diagrams and equations. There is a direct relationship between current and voltage. The greater the work to be accomplished, the more the current that is required. Most circuitry is designed to work within specific engineering requirements, with current and voltage established at maximum and minimum values. Exceeding either end of the scale will either cause the equipment to fail for lack of power or cause it to overheat and become damaged from overloading of the circuitry. Most electronic equipment today works at relatively low voltages that are converted from the standard 120 V AC to ±3 or 5 V DC within the equipment.

Voltage is generated from a number of sources. It is obtained by chemical reactions within a battery, from various forms of power-generating plants (e.g., coal, nuclear, hydroelectric), and from alternative energy sources such as the sun and wind. The amount of voltage available for use from a source is directly related to current and resistance. As current increases through a resistive material, the amount of force correspondingly increases, and the ability to do work also increases as well. The volt represents this value in a simple formula:

$$\text{Resistance} \times \text{current} = \text{voltage}$$

This is the basic relationship that is defined by Ohm's law. If one of the values in the equation is unknown, this law can be applied to any resistive circuit to get the unknown value. The importance of this law is evident when evaluating electrical demands for a communications room. Each device in a business environment—PBX, server, router, or switch—has a specific voltage and current requirement listed on its specifications sheet. What would happen if the current requirements on the circuit feeding the communications room exceed the value rating of the current in the electrical panel? Overload! An overload causes the circuit breaker to stop current flow to the circuit, as it is designed to do. By knowing the values and their relationships to one another, proper planning for system upgrades can be accomplished. This knowledge will allow one to understand the electrician when discussing physical plant requirements for facilities.

Ohm's law is always calculated using amps for current, volts for voltage, and ohms for resistance. Because these values can vary depending on the appliance or circuit (megavolts, milliamps, or kiloohms), they must be converted to satisfy the basic form of the value. Using the scientific notation chart in Exhibit 2.1 simplifies the conversion.

Capacitance

The ability of an electrical conductor to hold a charge is considered its capacitance. Capacitors are components in electronic equipment that are designed to accept and discharge electrical charges. They act as batteries within the circuitry, maintaining specific levels of voltage across designated components. When two oppositely charged conductors are placed in close proximity to each other, they create and hold an electrical charge until the power source of their charge is removed. The closer together the conductors are, the greater is the ability of the conductors to hold the charge. Capacitors take advantage of this property by separating two conductive elements with dielectric material that controls the charge–discharge rate. All wires within a cable bundle possess the ability to create conductance because of their close proximity to other cable pairs that are transmitting signals (which generates electrical charges). The insulating sheath of the wires helps reduce adverse conductance. However, if higher frequencies or power demands are pushed through the wires, the charge rate of the wires is significantly affected. Capacitance is directly related to frequency: as the frequency increases, the capacitance decreases, giving the circuit in which a coupling capacitor is situated an open condition to DC voltage. Exhibit 2.10 gives a basic graph of the relationship between frequency and capacitance.

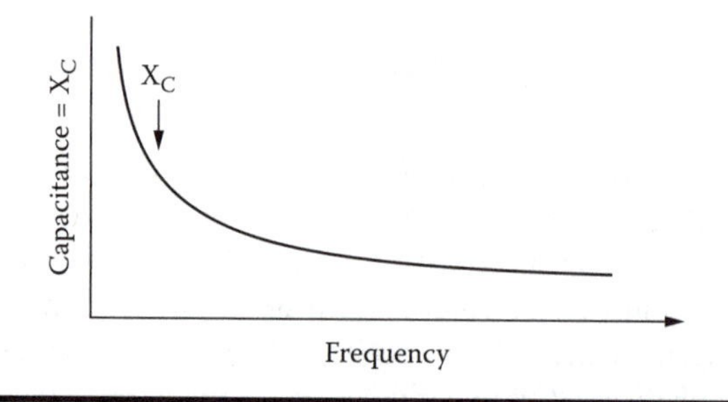

Exhibit 2.10 Relationship between capacitance and frequency.

Inductance

Wrapping wire around a core and passing current through it creates an inductor. The core can consist of numerous materials; however, magnetic material helps create an electromagnetic field around the core. The turns of wire and the strength of the core material magnetically increase the resistance of the inductor to a return to its precharged state. This property of an inductor is important in a number of components. Transformers, both step-up and step-down types, are built on the inductor principle. Coils, the devices used to send telephone signals over greater distances through the network, are designed as inductors. Inductance is the resistance of a conductor to a change of current direction. There is a small amount of inductance associated with cable pairs bundled together in a sheathed environment. It is negligible in most conditions; however, with increased frequency transmission through a pair of conductors, the inductive reactance can easily increase. Inductors act as filters to AC values while having little effect on DC signals. Exhibit 2.11 shows the relationship between frequency and inductance in a circuit. The difference between capacitance and inductance, and their relationship to frequency, is shown in Exhibit 2.12. The point at which the two resistive values meet is known as the *resonant frequency* of the components. Electrical circuit design engineers use the resonant frequency point to create filters that eliminate all frequencies but the

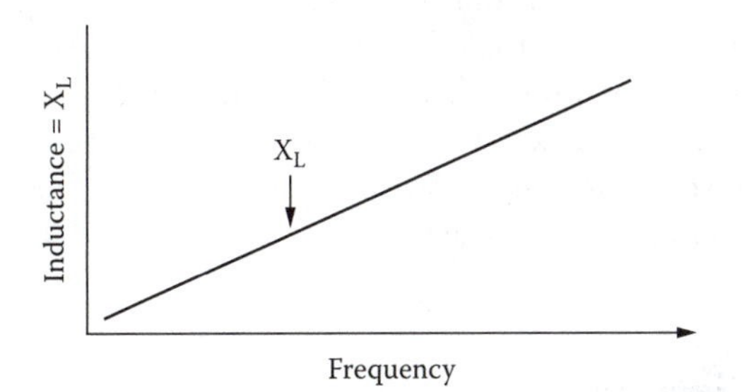

Exhibit 2.11 Relationship between inductance and frequency.

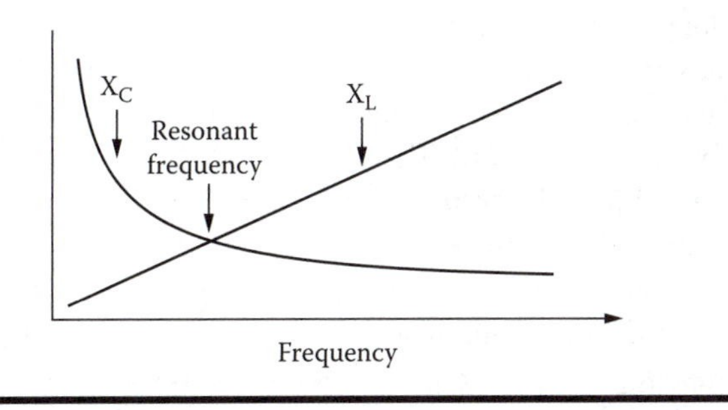

Exhibit 2.12 Resonant frequency.

resonant frequency. The filtering process is used to find a radio or television station and to clean up an analog circuit from extraneous noise due to higher frequencies before it is converted into a digital format.

Power

When electrons are moved between points of potential difference (e.g., positive and negative terminals on a battery), work is accomplished. The measure of the rate at which work can be accomplished is called power (P). The unit of measure used to define power is called the watt. A current flow of 1 A and an applied voltage of 1 V creates 1 W of power. Current, voltage, and power are related as follows:

$$\text{Current} \times \text{voltage} = \text{power}$$

As current increases, the amount of work that can be done also increases. The same holds true with an increase in voltage. However, simply increasing the voltage of a voltage source to try to gain more work out of an electrical motor will result in disaster. Each electrical device manufactured has a specific rating that dictates the maximum voltage and current drawn for that device. Most equipment has fuses to prevent excess voltage or current from being drawn; as a result, the device shuts down if an overload condition is detected. Electrical wires, connectors, and components are built to tolerate certain levels of power. It is not cost-effective to build devices that can work across all voltage and current values. Power is expressed in terms of watts, kilowatts, and megawatts.

Electrical Circuits

An electrical circuit is an interconnection of various electrical elements such as inductors, resistors, voltage sources, current sources, and transmission lines that form a closed loop, giving a return path for the current. Circuits may be analog or digital in nature. Circuits may be classified as linear or nonlinear. They may be constructed in series or in parallel, or a combination of both types.

When the various electrical components are connected one to one like a chain, then the circuit is said to be in series, as shown in Exhibit 2.13.

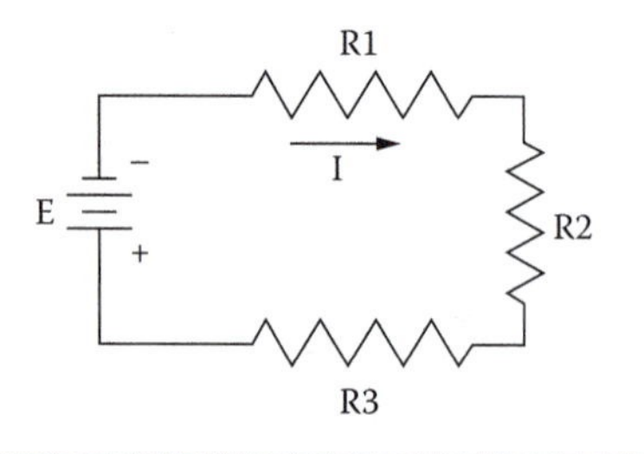

Exhibit 2.13 Simple series circuit.

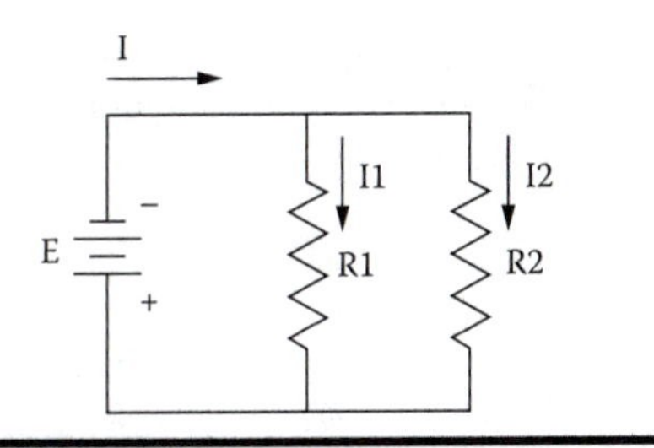

Exhibit 2.14 Simple parallel circuit.

A parallel circuit (Exhibit 2.14) is split into branches, and current flows through each of the branches simultaneously.

A complex circuit (Exhibit 2.15) is comprised of both series and parallel resistive components. The change of current across this type of circuit design is used to provide different power levels to different devices connected to the circuit.

A short circuit occurs whenever the resistance of the circuit drops to a very low value, most often to 0. Short circuits usually occur because of improper wiring and insulation (Exhibit 2.16).

Open circuits occur when there is a break in electrical connectivity. It can occur because of a break in the wiring or because of defective components in the circuit (Exhibit 2.16). The notation in Exhibit 2.16 provides a view of a complete circuit as a reference to open and shorted circuits.

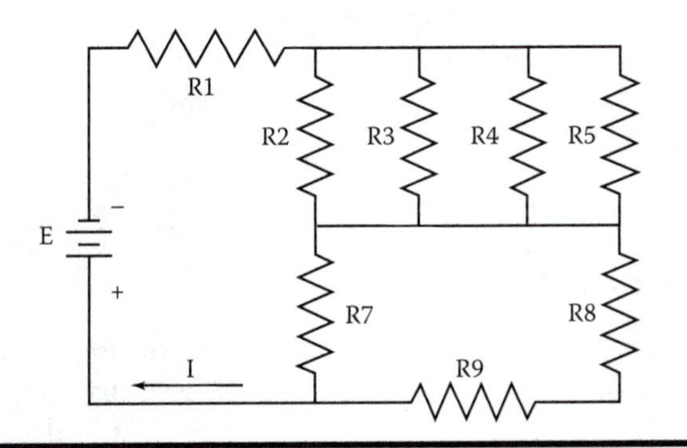

Exhibit 2.15 A complex circuit.

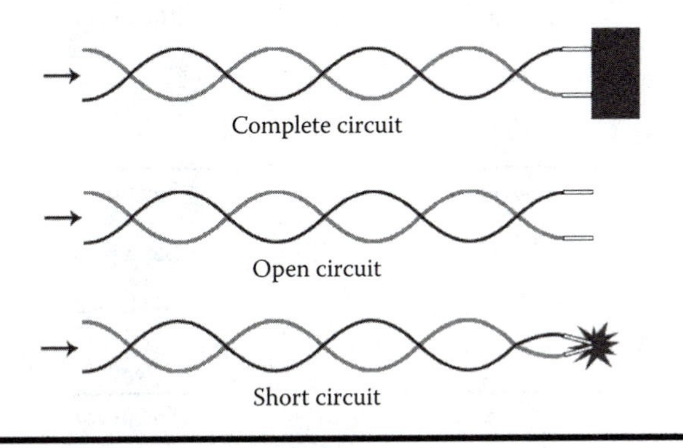

Exhibit 2.16 Types of electrical circuits.

The occurrence of either of the preceding two conditions (open or short circuit) results in an electrical malfunction, and the circuit needs to be repaired to restore normality.

Filters

Certain electrical appliances require one frequency or a specific range of frequencies from a band of frequencies for effective functioning. The circuit developed to perform this filtering process is called a filter circuit or simply a filter. There are several types of filters (Exhibit 2.17):

- Low-pass filters: these circuits filter out the high frequencies, allowing only the low-frequency components to pass through the circuit.
- High-pass filters: these circuits filter out the low-frequency components and allow only the higher frequencies to pass through.
- Band-pass filters: these filters allow only a specific range of frequencies to pass through from a mix of various frequencies.
- Band-stop filters: these filters reject a particular range of frequencies, allowing all other frequencies to pass through the circuit. They are also known as band-elimination or notch filters.

AC-to-DC Conversion

DC voltage is more efficient at doing work than AC voltage. However, the delivery of DC voltage over long distances, such as the grid that supports residential, commercial, and industrial power needs, is not feasible. On the other hand, AC is well suited for transport at high voltages and is stepped down to accommodate

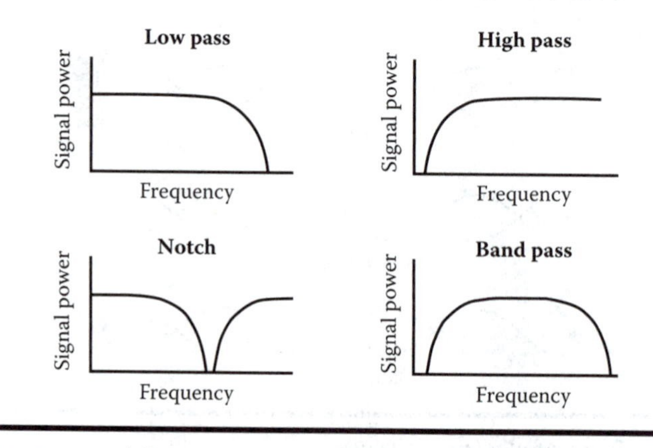

Exhibit 2.17 Types of filters.

the multiple-use requirements that it serves. Electronic devices, such as computers, televisions, routers, PBX, fax machines, and servers, work internally with DC voltage. Each device must convert the outlet 120 V AC supply to a DC value to make it operational for the myriad of chips, transistors, and other components running the device. DC is derived from AC by a device known as a *rectifier*, which uses special components to control the flow of AC current and bring the negative or alternating side of the signal together with the positive side of the signal. These devices, called *diodes*, allow current to flow through them in one direction only. There are numerous types of diodes, and they all perform a similar function that varies with voltage and current. The diode acts like a switch to current. Coupled with a capacitor, diodes can create a constant current flow for components within electronic equipment. Exhibit 2.18 depicts the first step in this conversion process, the normal sine wave signal from an AC source. The direction of the current flow is also shown. Exhibits 2.18 through 2.24 show the step-by-step conversion of AC to DC.

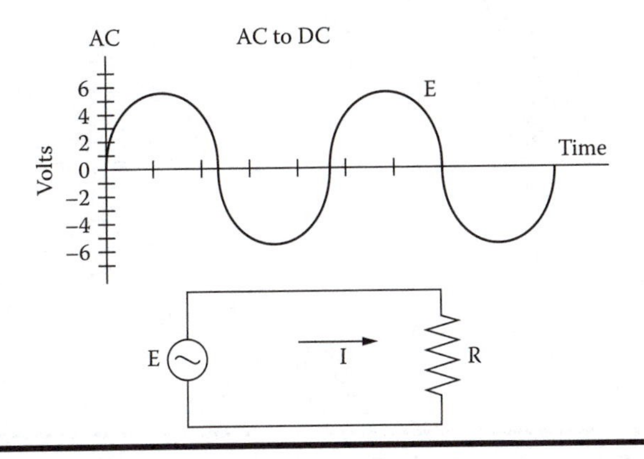

Exhibit 2.18 AC sine wave.

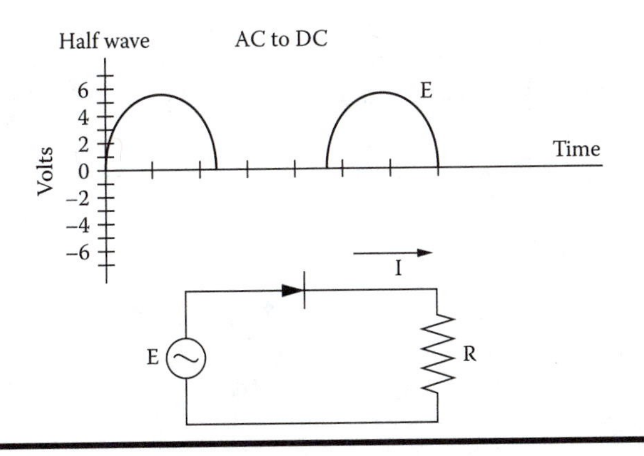

Exhibit 2.19 Diode eliminating negative current flow.

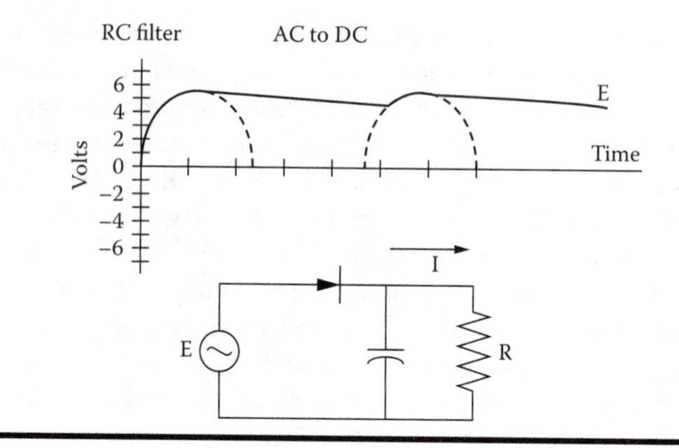

Exhibit 2.20 Capacitor added to an AC/DC converter.

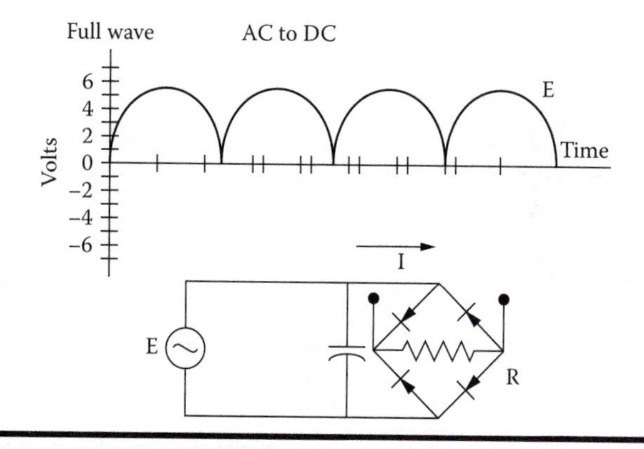

Exhibit 2.21 Full wave rectified AC signal.

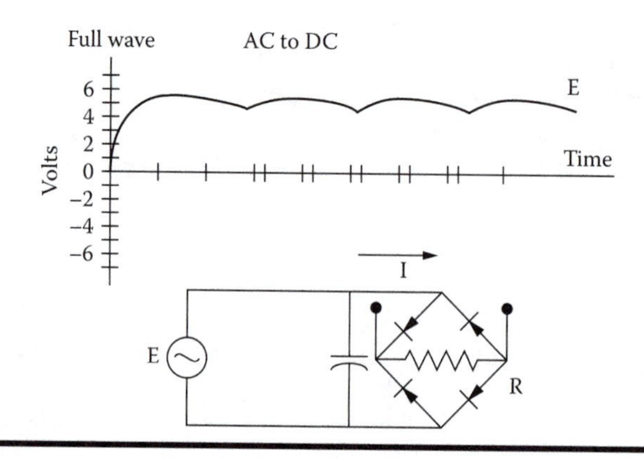

Exhibit 2.22 Full wave rectified with capacitor.

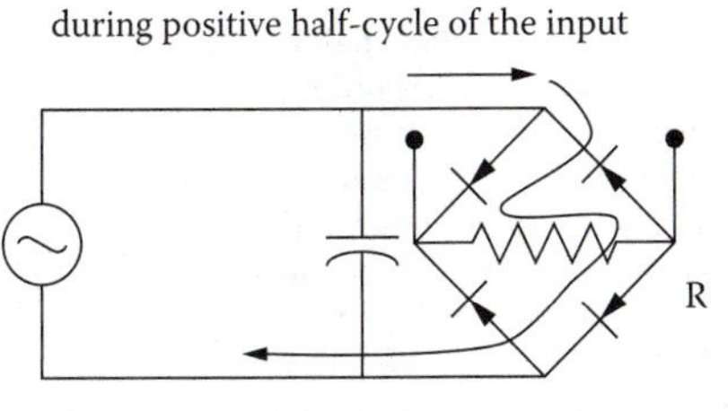

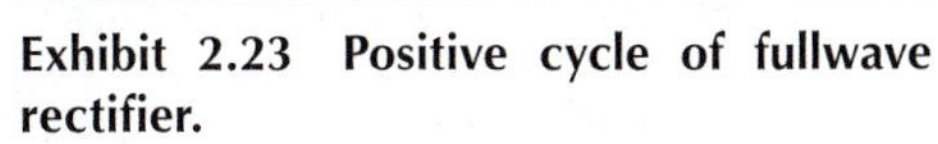

Exhibit 2.23 Positive cycle of fullwave rectifier.

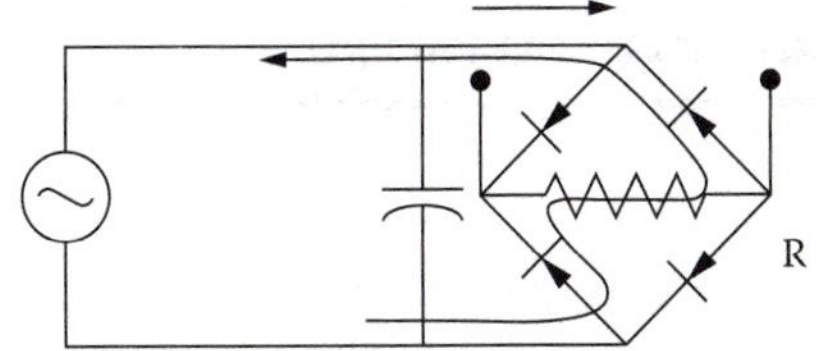

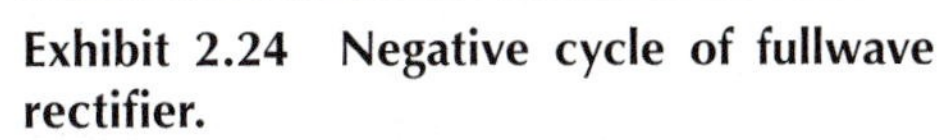

Exhibit 2.24 Negative cycle of fullwave rectifier.

A diode is put in place to prevent reverse current flow. Exhibit 2.19 shows this first step. Exhibit 2.20 shows a capacitor; when inserted into the circuit, it keeps the voltage from dropping back to 0. Remember that the capacitor acts like a battery in the circuit and discharges as the current is reversed in the circuit. This discharge process gives voltage back to the circuit to power it. We now need to bring the voltage to a more constant value and eliminate the drop that occurs prior to the charge sequence.

This is accomplished by creating a full-wave rectifier, which makes the negative current flow positive with the aid of diodes and a capacitor. By charging and discharging the capacitor at a higher speed, the attenuation of the voltage becomes negligible. Exhibit 2.21 shows how the sine wave is pulled from the negative side of the circuit by using rectifying diodes. Exhibit 2.22 shows how the capacitor charges and discharges to hold the voltage at a more consistent value. The voltage reading is taken at the two nodes on the rectifier.

Exhibits 2.23 and 2.24 show how the current flows through the rectifier during the positive and negative cycles. Again, the reading for the voltage is taken at the nodes on either side of the rectifier. Now that you understand the importance of DC voltage conversion, take a look inside your computer and follow the power lead from the external 120 V AC source through the unit. It is converted to DC immediately upon entry into the unit (Exhibit 2.25).

Summary

A basic knowledge of electricity is critical to understanding how the signaling process occurs across voice, data, and video networks. Knowledge of the correct

Developed as an alternative to hanging in the 1880s, the electric chair has been used to execute more than 4300 people. The electric circuit is rather simple: the condemned person's body is connected to a high-voltage circuit, becoming a resistance load. The electrodes are usually attached at the head and at the ankles. They are terminated by a sponge soaked in a saline solution to decrease the resistance in the circuit, the human body not being a good conductor, in general. There is no standard protocol for the voltage; however, if it is too low, it takes too long, and if it is too high, it burns the body. Nowadays, the voltage is usually set between 2000 and 2200 V for a current set between 7 and 12 A. Although in theory, the victim's nerves are immediately paralyzed, stopping all sensation of pain, the few who have survived execution said the pain was unbearable.

Exhibit 2.25 Case study: the electric chair.

nomenclature and scientific notation is important to understand all aspects of measurement and throughput discussed in communications technologies. The atom and its electrical properties were discussed to arrive at a better understanding of how current is created in an electrical circuit. Current, resistance, and voltage were explored as the three key components of electricity.

Ohm's law is the basic principle that shows how these three properties are related to one another. Power, inductance, and capacitance were discussed. Power is the measure of work performed within a circuit. Its unit, the watt, is pervasive in electricity and allied disciplines. Capacitance and inductance were discussed briefly in regard to their effect on electrical circuits and their ability to be coupled together to create a filter at specific frequencies. The conversion from AC to DC was examined. The importance of this conversion is underscored by every electronic piece of gear that needs to be plugged into an electrical outlet for power or to charge the batteries that provide the DC voltage for the circuitry to work.

Review Questions

Multiple Choice

1. What is the voltage provided by the central office (CO) to the telephone?
   a. 120 V AC
   b. 120 V DC
   c. 48 V AC
   d. 48 V DC
   e. None of the above

2. The waveforms vary in:
   a. Intensity
   b. Shape
   c. Duration
   d. Complexity
   e. All of the above
3. A signal is considered to be attenuating when:
   a. It is losing its intensity over time
   b. It is losing its period over time
   c. It is gaining its period over time
   d. It is gaining its intensity over time
   e. None of the above
4. A range of frequencies is called:
   a. Throughput
   b. Period
   c. Bandwidth
   d. Passband
   e. All of the above
5. What is Ohm's law?
   a. Resistance × voltage = current
   b. Voltage × current = power
   c. Current × resistance = voltage
   d. Current/voltage = resistance
   e. None of the above
6. As frequency increases:
   a. Capacitance increases and inductance increases
   b. Capacitance increases and inductance decreases
   c. Capacitance decreases and inductance decreases
   d. Capacitance decreases and inductance increases
   e. None of the above
7. The resonant frequency point is used to create filters that will eliminate:
   a. The resonant frequency
   b. All frequencies but the resonant frequency
   c. All frequencies higher than the resonant frequency
   d. All frequencies lower than the resonant frequency
   e. All of the above
8. What is the relationship between current, voltage, and power?
   a. Current × voltage = power
   b. Voltage/current = power
   c. Power × current = voltage
   d. Voltage × power = current
   e. None of the above

9. AC:
   a. Is less efficient at doing work than DC
   b. Is well suited for transport over long distances and at high voltages
   c. Is delivered at the wall outlet
   d. Must be converted to DC to serve electronic devices
   e. All of the above
10. A filtered full-wave rectifier is composed of:
   a. One diode and one capacitor
   b. One diode and one inductor
   c. Four diodes and one inductor
   d. Four diodes and one capacitor
   e. None of the above

Matching Questions

Match the following terms with the best answer.

Match the different values:

1. 1 V	a. 1000 V
2. 1 mV	b. 1000 mV
3. 1 mV	c. 0.001 mV
4. 1 kV	d. 0.000001 kV

Match the different values:

1. 280 Hz	a. 2800 kHz
2. 2.8 MHz	b. 28,000 Hz
3. 28 kHz	c. 0.28 kHz
4. 0.28 GHz	d. 280 MHz

Match the units:

1. Voltage	a. Ohm (W)
2. Current	b. Ampere (A)
3. Resistance	c. Watt (W)
4. Power	d. Volt (V)

Match the units:

1. Frequency	a. Second (s)
2. Period	b. Bit per second (bps)
3. Bandwidth	c. Hertz (Hz)
4. Throughput	d. Hertz (Hz)

Match the definitions:

1. Molecules	a. Do not possess an electrical charge
2. Protons	b. Are composed of atoms
3. Neutrons	c. Are negatively charged particles
4. Electrons	d. Are positively charged particles

Match the characteristics:

1. Copper	a. Conductor
2. Glass	b. Conductor
3. Plastic	c. Insulator
4. Gold	d. Insulator

Considering a simple electrical circuit (battery, wires, or load), match the values:

1. 10 V, 2 A	a. 5 W
2. 2 kV, 5 A	b. 25 kW
3. 0.5 kV, 20 mA	c. 5 kW
4. 25 mV, 5 mA	d. 400 W

Match the values:

1. 10 V, 2 A	a. 10 W
2. 2 kV, 5 mA	b. 10 mW
3. 0.5 kV, 20 mA	c. 200 mW
4. 25 mV, 8 A	d. 20 W

Match the definitions:

1. Voltage	a. The number of cycles divided by the time in which they occur
2. Current	b. The ability to do work
3. Power	c. The movement in the same direction in a conductor of the free electrons
4. Frequency	d. The measure of the rate at which work can be accomplished

Match the definitions:

1. Resistance	a. The ability of an electrical conductor to hold a charge
2. Capacitance	b. The force that opposes the flow of electrons in a material
3. Inductance	c. Allows current to flow only in one direction through it
4. Diode	d. The resistance of a conductor to allow current to change direction

Short Essay Questions

1. Define a waveform and name its four characteristics.
2. Define frequency and period, and name their respective units.
3. Define bandwidth and give an example.
4. Define current flow and name its unit. What are the differences between the two types of current?
5. Define resistance and name its unit.
6. Define voltage and name its unit.
7. Define capacitance.
8. Define inductance.
9. Define power and name its unit.
10. Describe AC-to-DC conversion briefly.

Chapter 3

Modulation Schemes

The first two chapters of this book have given you the fundamentals necessary to understand more complex topics associated with communications technologies. Modulation and multiplexing are the principles of moving information with progressively more efficient methods. We will use the knowledge gained in the previous chapters to help us understand specific information transfer methods. We will examine the radio-frequency spectrum, the modulation schemes that are used to maximize the limited resources of the spectrum, and new and emerging techniques of multiplexing that allow us to pack more information into a limited space on the spectrum. This examination will give us the information necessary to understand how radio stations broadcast and how cellular phones communicate through the airwaves.

Spectrum

The spectrum is the radio frequency available for personal, commercial, and military use. Specific portions of the spectrum are allocated for aircraft communications and navigation, commercial radio stations, broadcast television, military communications, and cellular telephony. The radio-frequency spectrum is a limited resource; there are only so many available licenses that can be issued for use of the spectrum. The recent auctions (and reauctions) by the Federal Communications Commission (FCC), the federal regulatory body responsible for spectrum allocation in the United States, are an indication of how highly valued the available spectrum is. Some companies have bid so high for the licensing rights to a specific spectrum that they became bankrupt from those costs before they were able to implement the infrastructure for service delivery.

Exhibit 3.1 gives a reference to various bandwidths of frequency and the service they are currently assigned to provide. Exhibit 3.2 provides a graphic representation of the various devices and where they can be found on the electromagnetic spectrum.

Each national governing body controls the use of the spectrum. The International Telecommunications Union (ITU) defines international standards, regulates international radio frequency, and facilitates the development of communications technologies worldwide. Within the ITU, the radio communication sector (ITU-R) is responsible for the management of the radio-frequency spectrum and satellite orbits. Every 2 to 3 years, the World Radiocommunication Conference is held to review and revise radio regulations when necessary. Because there is a limited amount of spectrum available for commercial use, this global governing body serves an important role in harmonizing the various demands on the spectrum.

The spectrum in the United States is allocated in three categories: government, licensed, and unlicensed. Government spectrum allocation covers military, navigation, and secure communications, and aviation and public-safety frequencies. The licensed spectrum includes cellular technologies, wireless local loop, satellites, and radio and television stations. The unlicensed spectrum that is available for use in the United States is from 902 to 928 MHz, 2.4 GHz, and 5.8 GHz. These bands are also known as the industrial, scientific, and medical (ISM) frequencies. The

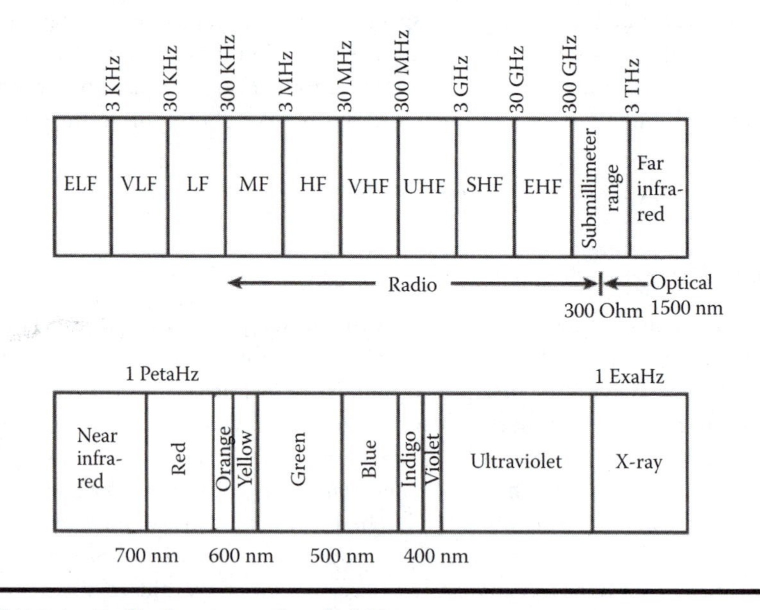

Exhibit 3.1 Radio-frequency bandwidths.

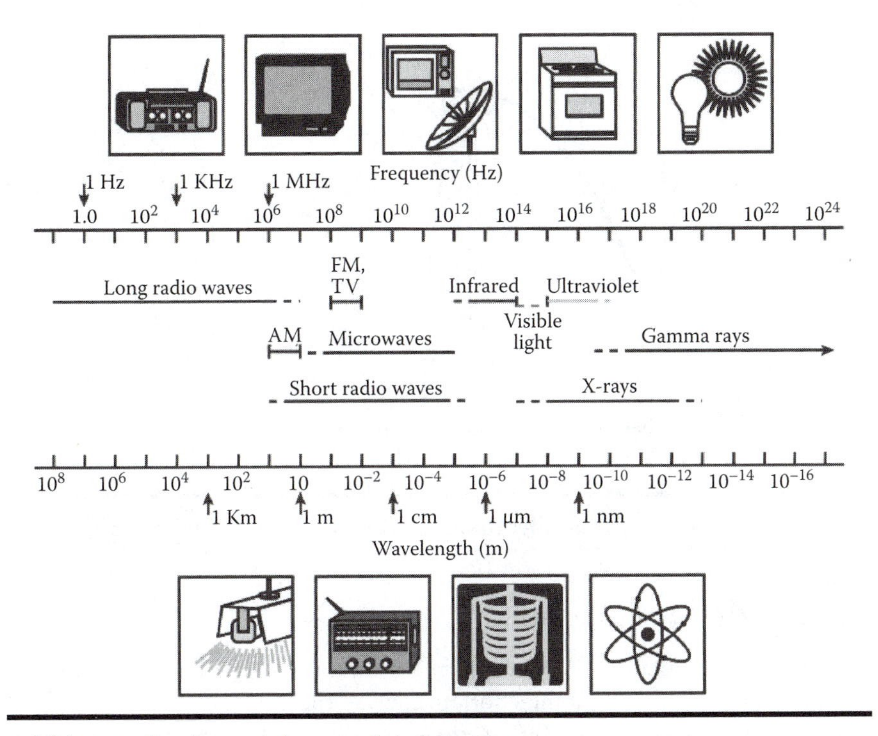

Exhibit 3.2 Device and the associated spectrum.

FCC's website provides a full listing of every spectrum allocation currently distributed (http://www.ntia.doc.gov/osmhome/allochrt.pdf).

As we move up the spectrum, radio waves give way to light waves. This indicates that as wavelengths become smaller, they become susceptible to conditions that affect the transmission of light. Frequencies below 6 GHz are not greatly affected by line-of-sight issues or obstructions blocking the signals. Beyond 6 GHz, we are bound by the limitations associated with the curvature of the earth, atmospheric conditions such as fog and rain, and buildings and other structures that impede the signal's transmission path. We will discuss more of these issues as we delve into the specific wireless technologies in this and subsequent chapters.

The goal of information transfer is to get the greatest possible throughput within the bandwidth allocated for that specific service. Increasing the throughput either requires an increase in bandwidth, thus obtaining more portions of the limited spectrum, or requires more information to be transmitted using the existing bandwidth, using modulation schemes. Exhibit 3.3 shows the bandwidth required for different communications technologies that use radio waves to deliver information.

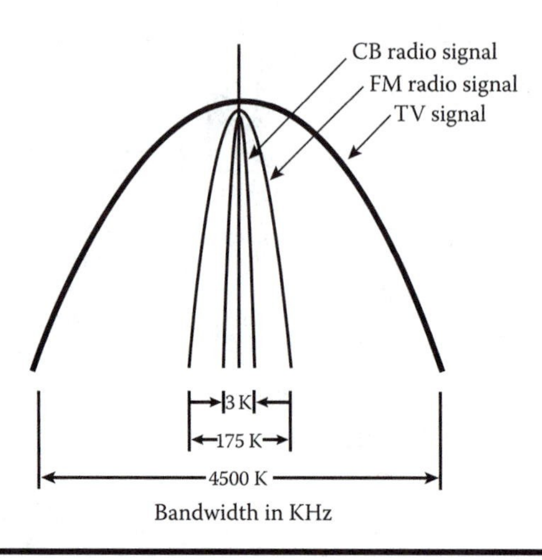

Exhibit 3.3 Throughput considerations for communications technologies.

Amplitude Modulation

In amplitude-modulation (AM) frequencies, the carrier signal adds to an information signal to create a joint effort to deliver the original signal as far and as efficiently as possible. Exhibit 3.4 shows a theoretical perspective of what an information signal and carrier signal may look like. The cycles-per-second graphical comparison in Exhibit 3.4 shows the difference in frequency between a signal and a carrier. At AM frequencies, the signal is used to excite or modulate the

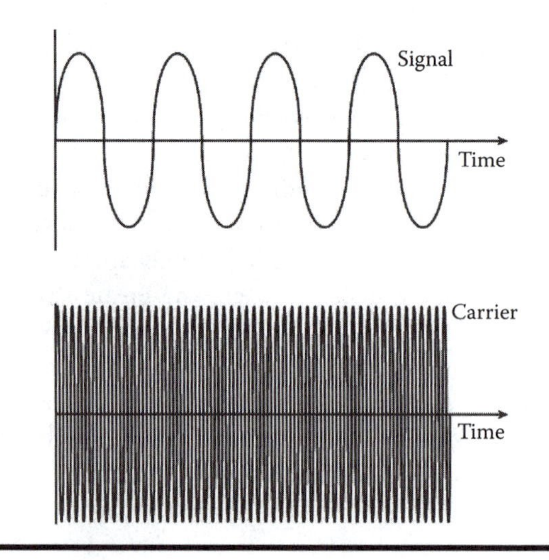

Exhibit 3.4 Comparison of signal and carrier waveforms.

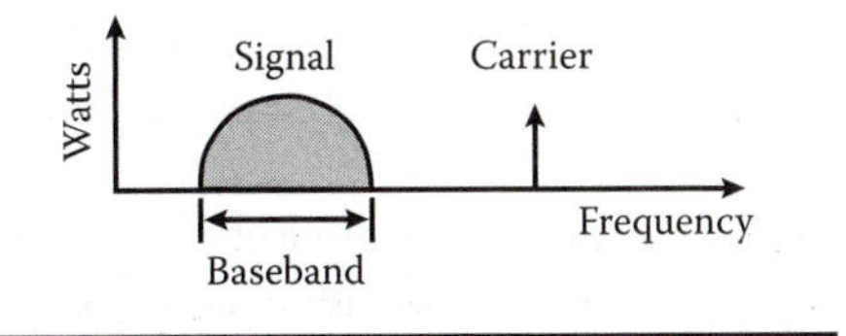

Exhibit 3.5 Perspective of signal and carrier differences in relation to frequency.

carrier and controls the amplitude of the signal, etching the information onto the carrier signal in an additive type of process. Exhibit 3.5 shows another perspective on the spectrum continuum of the difference between the baseband (signal) frequency and the carrier.

By placing the signal onto the carrier (added using a nonlinear method), we have the ability to move information with electromagnetic waves (radio waves) over incredible distances. Exhibit 3.6 shows a more accurate perspective of what a voice signal may look like before modulating an associated carrier frequency. Exhibit 3.7 shows the result of the carrier frequency being modulated by the signal, creating an information signal based on the amplitude of the carrier frequency. The carrier frequency reaches a maximum

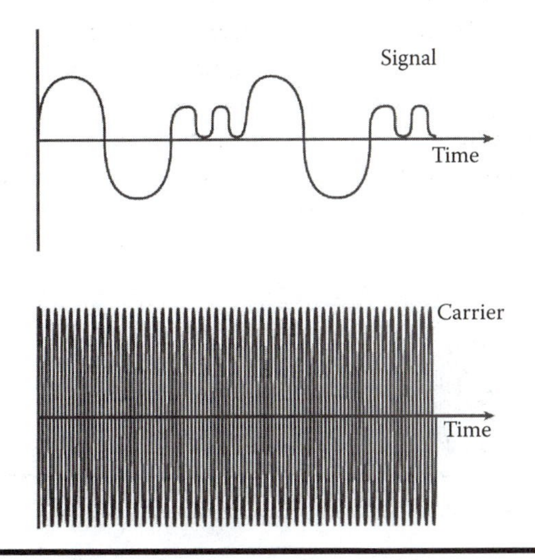

Exhibit 3.6 Representative signal and carrier frequencies before modulation.

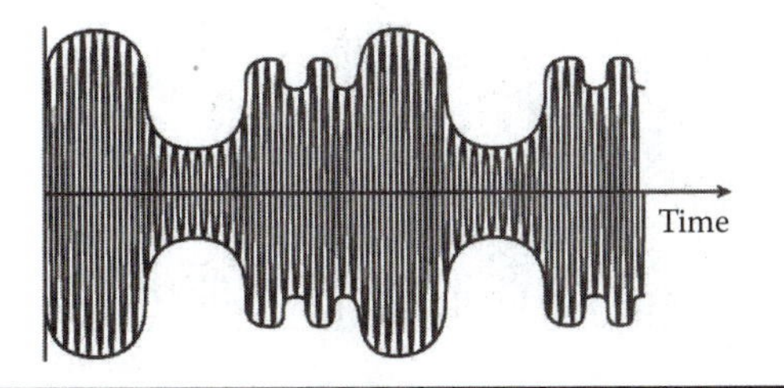

Exhibit 3.7 Amplitude-modulated carrier frequency.

value corresponding to the highest positive value of the information signal. The carrier also reaches its minimum value when the information signal is at its lowest negative level.

We can see the original information signal traced on the amplitude of the carrier frequency. We can also see a mirror image of the information signal on the negative side of the carrier signal. This envelope of information can be reduced to a single side of information by filtering out the lower side of the signal. This process creates greater bandwidth usage and is called a single-sideband carrier (SSB). Exhibit 3.8 shows an example of an SSB signal.

AM is highly susceptible to noise. Any high-power electromagnetic interference that is present in the atmosphere can add to the AM signal. This added power raises the amplitude of the signal beyond a recognizable range; it is the reason we hear the static condition at the receiving end of the transmission when listening to an AM radio station during electrical storms. Exhibit 3.9 represents noise added to an AM signal. Note the spikes of power above and below the signal envelope. These power surges are detected as noise by the receiving end of the signal. The efficiency and simplicity of an AM transmission make it possible to implement it throughout the country in stationary and mobile configurations. Although surpassed by other modulation techniques for quality of sound applications, AM continues to be a viable technology for delivering information. A simple block diagram provided in Exhibit 3.10 helps us understand how information travels, modulates the carrier, and reaches the antenna for broadcasting.

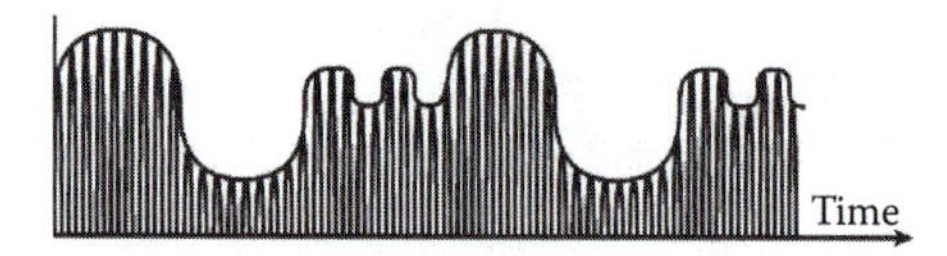

Exhibit 3.8 Single-sideband carrier (SSB) frequency signal.

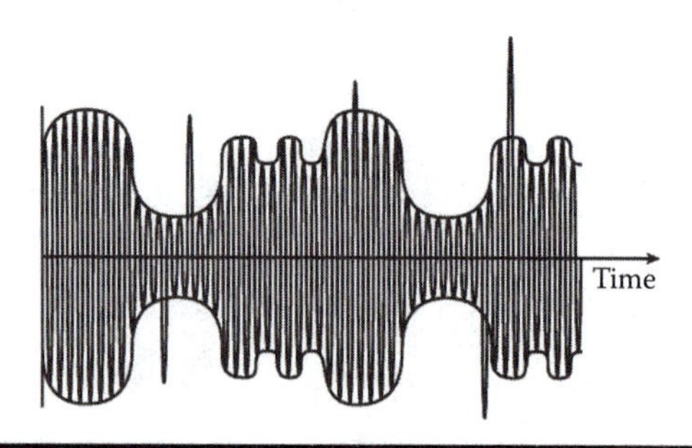

Exhibit 3.9 Electromagnetic noise and an AM signal.

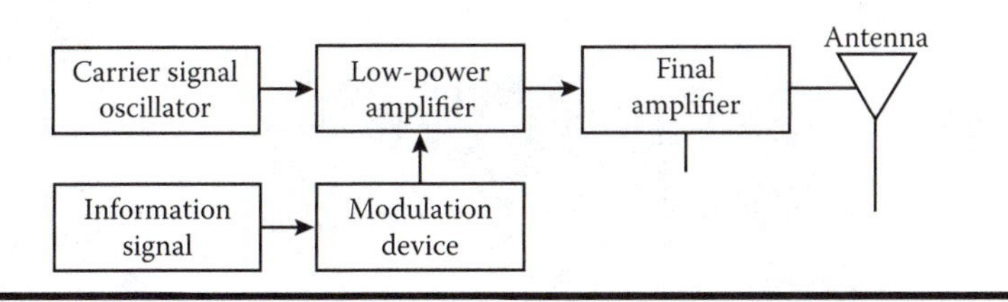

Exhibit 3.10 Simple block diagram of amplitude modulation.

Frequency Modulation

At frequency modulation (FM) frequencies, the information signal causes the carrier signal to increase or decrease its frequency based on the waveform of the information signal. The rate and amount of frequency change are related to the frequency and amplitude of the input signal. Exhibit 3.11 gives an example of how the information signal modifies carrier frequency behavior.

The carrier signal's frequency is altered with respect to the amplitude of the information signal. At higher amplitudes in the information signal, the frequency is higher in the carrier signal. At lower amplitudes in the information signal, the frequency is lower in the carrier signal. Exhibit 3.12 shows how the carrier frequency is affected by the information signal.

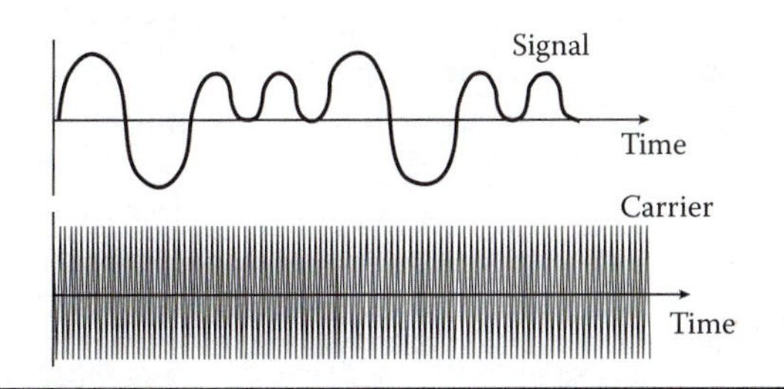

Exhibit 3.11 Information signal and carrier frequency for frequency modulation.

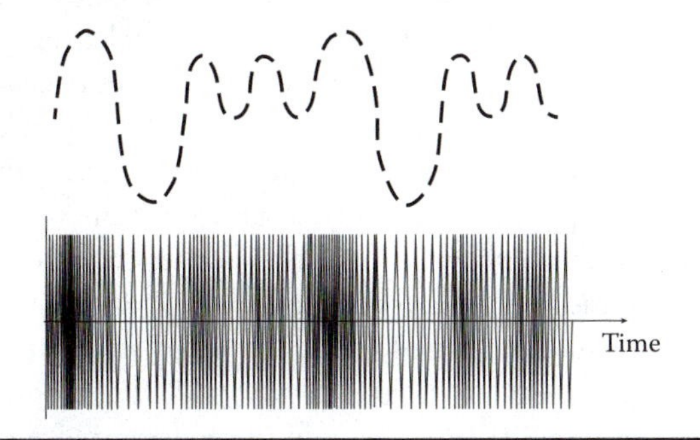

Exhibit 3.12 Frequency-modulated carrier signal.

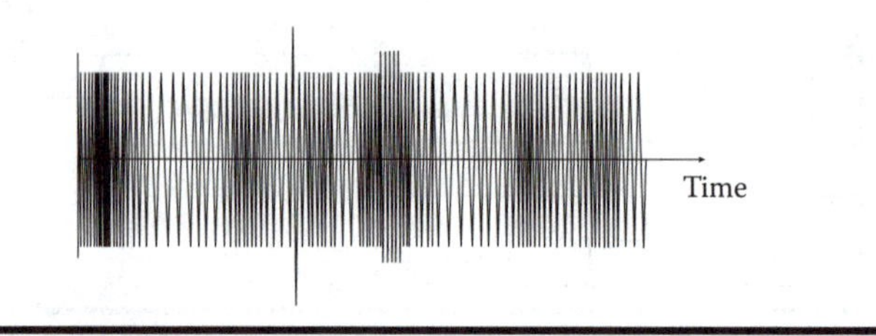

Exhibit 3.13 FM signal noise.

As the diagram shows, at the peaks of power from the signal, the carrier frequency is increased to its maximum rate, which is depicted by the closer widths of the cycles. As the amplitude swings to a lower power, the frequency decreases in speed.

The greatest advantage of FM signaling over AM is that it is less susceptible to noise. Because the amplitude of the FM signal is not affected by the signal source that modulates the carrier, adverse amplitude changes have little impact on the transmission quality. Any spikes added to the amplitude of the FM signal are filtered off on the receiving end of the transmission. Exhibit 3.13 represents noise added to an FM signal.

The downside of an FM transmission is that it does not have the same range as an AM transmission. FM carrier frequencies are three orders of magnitude higher than those of AM carriers (kilohertz for AM as opposed to megahertz for FM). As we move up the frequency spectrum, signals start to exhibit characteristics of light waves. Although FM frequencies are still a distance away from actual light frequencies, issues such as the curvature of the earth and geographic obstructions still affect them.

Phase Modulation

When an information signal is examined, it looks similar to a sine wave from trigonometric functions. Because a sine wave is defined by degrees ranging from 0° to 360°, we can think of the sine wave as a complete circle; the cycle of a wave is defined by when it repeats its form over a specific period of time. If we could start an information signal of the same amplitude and frequency at differing phases, change could be detected by the receiving equipment based on the starting phase shift, and we could incorporate more information within the same frequency spectrum based on the different phase shift of the frequency delivered. Exhibit 3.14 shows how the information signal can be shifted by 90° angles.

New information could be started on our depicted signal at the 0°, 90°, 180°, and 270° phase shifts. There are various forms of coding associated with phase modulation. Phase-shift keying (PSK) is used in modems, voice coders, and other transmission equipment that is limited to the frequency spectrum but has a need

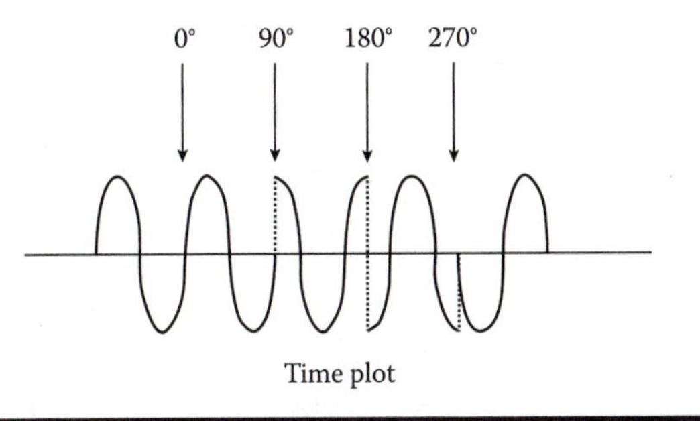

Exhibit 3.14 Phase modulation with a 90° phase shift.

for increased throughput. By using these creative methods of maximizing allocated frequency, or frequency capacity limited by Shannon's law, we can increase the throughput of numerous technologies.

Pulse-Amplitude Modulation

Pulse-amplitude modulation (PAM) is the first step in converting analog waveforms into digital signals for transmission. PAM was used extensively in the all-electronic generation of telephone switching gear. Sampling the signal waveform and ascribing a voltage value to the point on the wave where it was sampled defines PAM. Exhibit 3.15 depicts a signal being sampled (vertical lines) at a very low rate. Exhibit 3.16 shows what the resulting voltage amplitudes would look like once analog information is defined in electronic form.

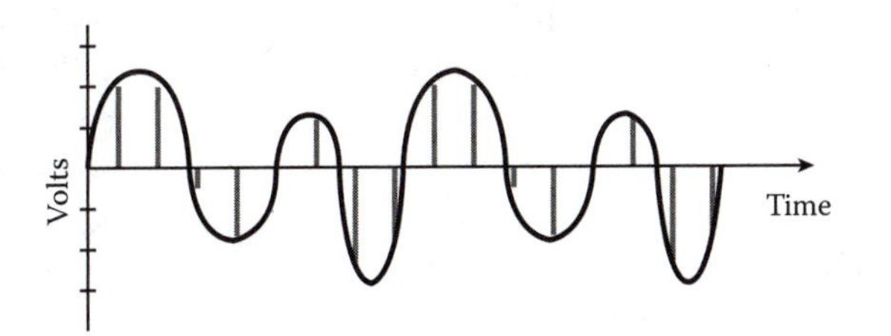

Exhibit 3.15 Analog waveform sampling.

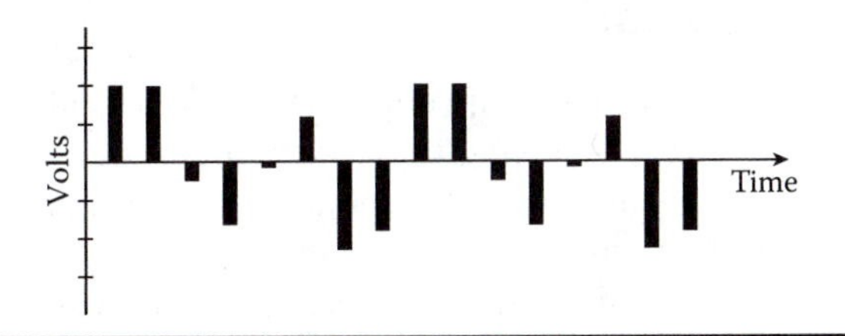

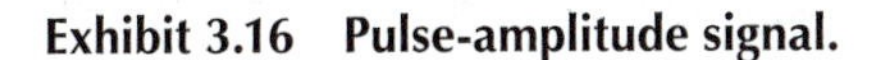

Exhibit 3.16 Pulse-amplitude signal.

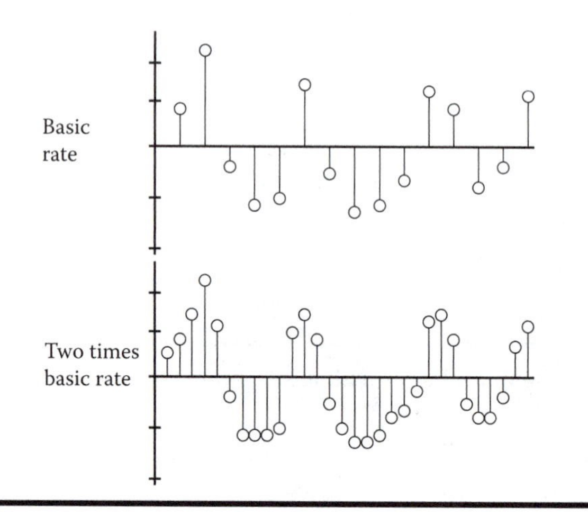

Exhibit 3.17 Nyquist sampling examples.

The sampling rate of the analog wave defines the quality of the signal. Increased sampling helps replicate the signal closer to its original form. Nyquist's theorem dictates that sampling should occur at a rate that is twice the highest frequency being sampled. The human voice travels a range from approximately 30 to 24,000 Hz. However, most of the information being transmitted resides in the range from roughly 300 to 3700 Hz. Engineers created a filtering process that would allow the sampling of a bandwidth of 4000 Hz (4 kHz), which would encompass the frequencies defined. The sampling rate, needing to be twice the highest frequency being sampled (4 kHz), would result in a sampling rate of 8 kHz. To attempt to image sampling something at such a high rate in a very short period of time always leaves me in awe of how digital systems were originally designed. Exhibit 3.17 compares a basic rate to twice the basic sampling rate and the resulting replicated waveforms. The exhibit showing twice the basic rate can be seen to have a greater resemblance to the original waveform based on the points of each pulse. Defining those pulses and their associated voltage values in a binary format is the next step in moving to digital transmission technologies.

Pulse-Code Modulation

Because we have defined the process of sampling an analog signal, and how we need to provide for a high rate of sampling per second to provide for a close facsimile of the original signal, we can move on to defining how we digitize this electronic information. Pulse-code modulation (PCM) provides a process in which each PAM signal is converted into an 8-bit binary character. The binary format of using just 1s and 0s creates a condition of either on or off, voltage or no voltage, and light or darkness. Because we can represent any number in the decimal system with a binary format, we can use just two signals to deliver any combination of numbers.

Representation of an infinite range of numbers by transmitting just two different states in various configurations is the basis of digital transmission. By providing 8 bits (1 byte) for each PAM signal, we now need to create a way to deliver this information in an orderly fashion, to keep all the 1s and 0s from running into one another. Exhibit 3.18 shows an example of the conversion from PAM to PCM.

The binary numbers chosen here are arbitrary but represent an actual conversion of the PAM signal. A more sophisticated process occurs in quantizing the information. Quantizing is a systematic method of providing standard binary numbering to PAM samples for PCM conversion. Nested within quantizing is the procedure called *companding.*

Companding is the process in which a greater number of samples are provided at lower-power conditions of the signal waveform rather than at the higher-power portions of the same waveform. This is done to reduce the signal-to-noise ratio (Shannon's capacity law, again) at points along the signal where the ratio is such that noise may have a greater impact on the quality of the signal. The word *companding* was obtained by combining the words *compress* and *expand.* Networking equipment will compress the sending end signal as described earlier, whereas the receiving equipment will expand the compressed information back into a recognizable waveform. Exhibit 3.19 depicts the assignment of more values to the lower-power sections of the information signal. Note that at both the positive and negative portions of the waveform at the low-power areas, there are more values assigned for ascribing values to PAM samples.

After this portion of the digitization process is completed, the binary values are encoded into a format that is recognizable at the receiving end of the digital signal. All of these 1s and 0s stay organized because of a process known as *multiplexing,* which we will cover in Chapter 4.

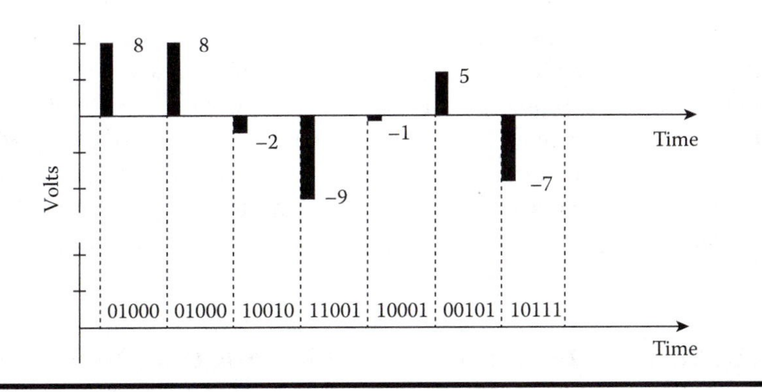

Exhibit 3.18 Converting from PAM to PCM.

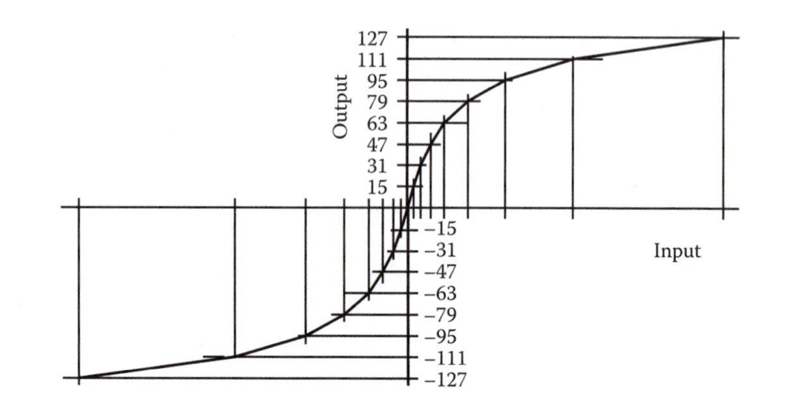

Exhibit 3.19 Companding.

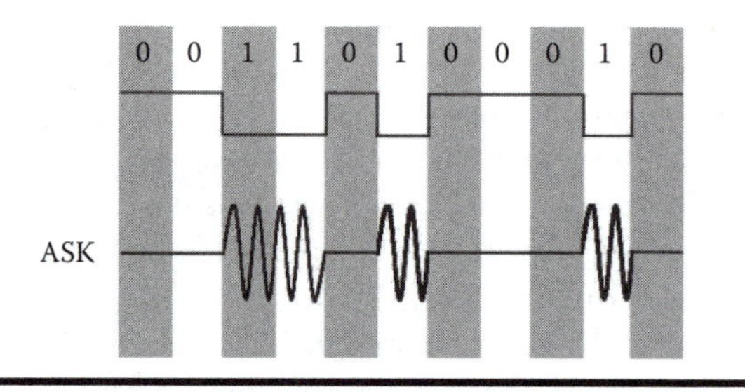

Exhibit 3.20 Amplitude shift keying.

Amplitude Shift Keying

Amplitude shift keying (ASK) uses the basic form of an analog AM delivery system. The ASK technique overlays digital information onto the carrier frequency instead of the normal analog information that would be put on the carrier frequency. The transmission is sent through high- and low-amplitude signals. The high-amplitude signal is replaced by a binary 1 (1), and the low-amplitude signal is replaced by a binary 0 (Exhibit 3.20). Because there is a big difference between high- and low-amplitude signals, this replacement greatly reduces the chance of the signal being misread. This is important because AM is normally more susceptible to noise and energy spikes in the frequency. The downside to this method is the susceptibility to sudden gain changes. Also, it is an inefficient modulation technique.

Adaptive Differential Pulse-Code Modulation

To understand adaptive differential pulse-code modulation (ADPCM), it is best to start out talking about the *DPCM* part of the acronym. PCM was discussed earlier in the chapter, and the differences between the input sample signals are small enough so that instead of transmitting the entire sample signal, it transmits the difference. Because this difference is smaller than the entire signal, the number of bits transmitted is reduced to around 48 kbps instead of the normal 64 kbps used with PCM by itself.

To further explain this process, the input signal is sampled at a constant sampling frequency, which is twice the input frequency (i.e., *Nyquist's rule*). The sample, using the PAM process, is then processed into binary values. The sampled input signal is now stored in what is known as the *predictor*. The predictor has to take the stored sample signal and put it through a *differentiator*, which then compares the previous sample signal with the current sample signal. The difference is put through the quantizing and coding part of the normal PCM. The difference signal is now sent to the final destination, where everything will be reversed to reconstruct the original input signal. There are problems with DPCM in regard to voice quality because uniform quantization is better with higher signals when the human voice usually generates a smaller signal. This is where the adaptive part of ADPCM comes in.

ADPCM will adapt the quantization levels of the difference signal that are generated in the DPCM process. It does this in two ways: if the difference signal is low, it will increase the size of quantization levels, and if the difference signal is high, it will decrease the size of quantization. This method further reduces the bit rate of voice transmission down to half of what the normal modulation scheme requires, that is, from 64 to 32 kbps.

Phase-Shift Keying

PSK relies on the carrier switching between two phases of the signal to define the binary status of information being sent. The binary status of 0 is represented with a signal burst of the same phase as the previous signal burst. The binary status of 1 is represented with a signal burst of opposite phase to the previous signal burst. It is a very efficient data delivery method owing to the low bit error rates of delivery. There are a few different PSK methods used in wireless networking systems. They are the binary PSK (BPSK) method, differential PSK method, and quadrature PSK (QPSK) method.

The least complex method is the BPSK method (Exhibit 3.21). BPSK uses a change, or shift, in the normal cycle of the carrier from the base of 0° to 180° from that initial signal phase. These changes to the signal will define the binary condition without changing the amplitude or frequency of the carrier signal.

The next more complex method is the differential PSK method (Exhibit 3.22). This method is always implemented with reference to the previous bit of information that was transmitted. If the previous bit was a binary 0 and the next bit is also a 0, there occurs a signal burst of the same phase as the previous signal burst to represent the new digit of the same amount. If the new bit changes and is a binary

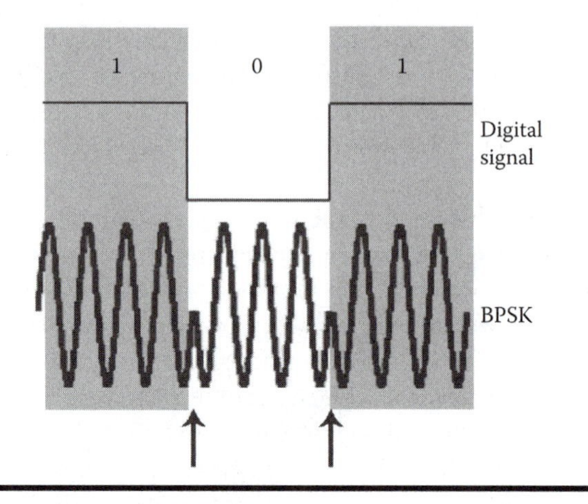

Exhibit 3.21 Binary phase-shift keying.

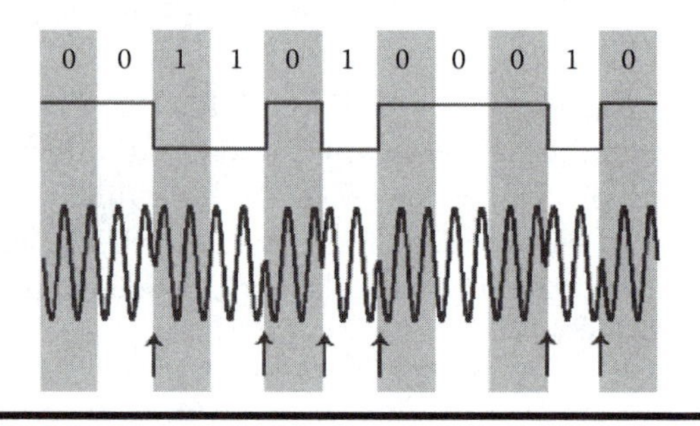

Exhibit 3.22 Differential phase-shift keying.

1, then a signal burst of the opposite phase (stopping and then restarting the wave in the opposite phase) to the previous signal burst occurs. Therefore, the previous phase angle determines the binary value of the next shift.

The last, and even more advanced, method is the QPSK method. The goal of this method is to transmit more information within the defined frequency allocation. This method uses several phase offsets and will provide more than 1 bit of information per frequency shift. Because multiple bits are used, there is greater throughput. The four-phase-shift method will produce 2 bits, and the eight-phase-shift method will produce 3 bits. The modulated wave shifts for the four-phase-shift method lie between four phases that are each 90° apart.

Orthogonal Frequency-Division Multiplexing

Orthogonal frequency-division multiplexing (OFDM), also known as multicarrier modulation, is an advanced coding scheme used by mature and emerging wireless and wire-line technologies. Wi-Fi, various forms of digital subscriber line (DSL) services, and fourth-generation (4G) broadband mobile wireless technologies such as 802.16 (WiMAX) and long-term evolution (LTE) all incorporate OFDM into their coding schemes. OFDM uses channels (frequencies) that do not overlap and are each defined in bandwidth based on the protocol that is being used (e.g., 802.11a). These channels carry bits of the original information (or data) stream, similar to frequency-division modulation (FDM); however, each stream or sub-channel is obtained from a single data source. Depending on the protocol, the sub-channel allocation varies by available spectrum bandwidth. For example, 802.11a uses 54 channels, while WiMAX uses 192 channels!

There are a number of advantages to OFDM that make it viable for the modulation requirements of communications technologies. Because the information can be spread across numerous subcarriers, interference in a single-frequency space is mitigated. Dynamically assigning throughput capabilities associated with the subcarriers provides for higher efficiency of data delivery, and the spectrum does not remain idle if needed by other subscribers. The major advantage of OFDM is its capability of putting more information into parallel (as opposed to serial) data streams while limiting the interference between and within channels, adding greater data delivery within the limited frequency spectrum.

The demand moving forward in the mobile environment for greater data loads will require modulation methods to allow high-bit-rate applications to work anywhere, anytime. As portions of spectrum are auctioned off, service providers will find it necessary to utilize a modulation scheme that can adapt to a noncontiguous spectrum. OFDM will be relied on heavily as migration to 5G mobile communications, software-defined radio, and time-division duplexing on a single frequency follow user demand for high data rates.

Quadrature Amplitude Modulation

Quadrature amplitude modulation (QAM) is basically a combination of ASK and PSK. Two different carrier signals are sent simultaneously on the same frequency. When talking about the wave and how it shifts, we can start out by simply assigning a 3-bit value. Those 3-bit values can make eight different combinations: 100, 010, 001, 110, 111, 000, 011, and 101. Each of these eight combinations is assigned an amplitude and a phase shift (none, 1/4, 1/2, 3/4). To understand this concept better, look at Exhibit 3.23. As you can see, when the wave goes through the point, it is represented as one of the eight combinations of bits. To send more information through, you just increase the number of combinations. For example, in 16 QAM, 16 combinations are possible, with each point being assigned a 4-bit value (explaining the diagram in my own words). Currently, with wireless networks, the size of QAM is pushing 54 Mbps throughput with the QAM configuration at 64.

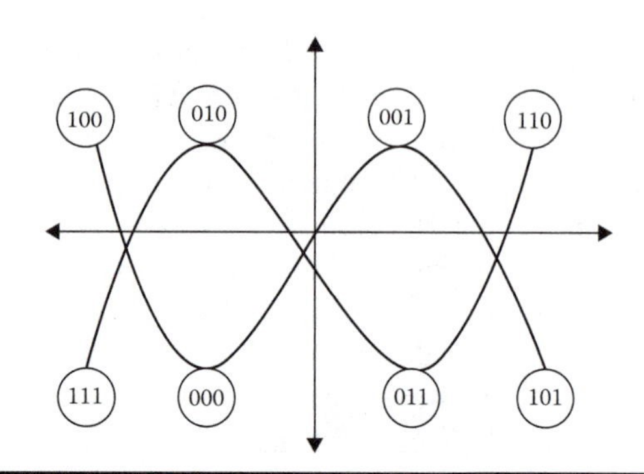

Exhibit 3.23 An example of QAM coding points.

Multiple-Frequency Shift Keying

Multiple-frequency shift keying uses two or more frequencies. What happens is that a signal is sent at a certain frequency for a fraction of a second. After that time period of a fraction of a second expires, the frequency can be changed. The frequency at a certain time determines the bits of information sent at that time. A method to help ensure that there is no data corruption is forward error correction. In the most basic form of forward error correction, each bit is sent twice and checked with protocol. The bit closest to protocol is accepted. For this concept to work, the receiver of the frequency needs to be able to hold a constant frequency for long periods of time so that there are no interruptions in the data transmission. The most common number of frequencies used in this method is 16. Each frequency represents a different set of information.

Summary

Modulation schemes for delivering information from a source to a receiver have evolved from simplistic delivery systems to extremely complex and sophisticated methods. The foundational material presented in this chapter should provide the reader with the tools to understand how information is moved across various mediums and the positive and negative attributes associated with each scheme.

AM and FM have been used for numerous delivery systems ranging from radio station broadcasts to cellular technologies. Phase modulation contributes a more sophisticated way of bringing additional information onto the existing frequency without increasing the allocated bandwidth.

PAM was the first step in converting analog signals into the digital environment. From PAM, we moved into PCM.

The sophistication of PCM is evident by the complexity of the Nyquist sampling process, quantizing, companding, and encoding. How PCM is implemented in networks so that various technologies can talk to one another will be covered in greater detail in a subsequent section on multiplexing.

Review Questions

True/False

1. In basic phase-shift keying, the amplitude of the signal is changed when transmitting information.
   a. True
   b. False
2. In amplitude modulation, the signal is used to increase or decrease the frequency of the carrier wave.
   a. True
   b. False
3. According to Nyquist's theorem, a frequency with a high point of 16 kHz should have a sampling rate of 32 kHz.
   a. True
   b. False
4. PAM is the first step in converting analog waveforms into digital signals.
   a. True
   b. False
5. Companding is the process in which a greater number of samples are provided at the higher-power conditions of the signal waveform.
   a. True
   b. False

Multiple Choice

1. Multiplexing is the process of:
   a. Using technological methods to put more information into limited bandwidth
   b. Using a carrier signal to deliver information transfer more efficiently
   c. Assigning frequencies to various organizations to achieve information transfer
2. The international regulating body concerned with frequency allocation:
   a. Federal Communications Commission (FCC)
   b. World Administrative Radio Conference (WARC)
   c. International Telecommunications Union (ITU)
3. Government allocation spectrum:
   a. Television, radio, and satellite
   b. WLANs, cordless phones, and microwave ovens
   c. Military, aviation, navigation, and secure communications

4. Unlicensed spectrum allocation:
    a. WLANs, cordless phones, and microwaves ovens
    b. Television, radio, and satellite
    c. Military, aviation, navigation, and secure communications
5. Transmitting signals above the 6 GHz range can be limited by:
    a. Interference caused by power spikes
    b. Interference cause by household appliances and cordless phones
    c. The curvature of the earth, atmospheric conditions such as fog and rain, or buildings that might block the path of transmission
6. What is modulated in AM?
    a. Frequency
    b. Power
    c. Amplitude
7. The information carrier in AM reaches its minimum value when the information signal is at its:
    a. Highest value
    b. Lowest value
    c. Average value
8. Which of the following is not a benefit of OFDM?
    a. Interference mitigation
    b. Channel overlap
    c. Dynamic assignment of throughput
9. AM radio is very susceptible to noise; this is caused by:
    a. High-powered electromagnetic interference, which raises the amplitude of the information signal to an unacceptable level
    b. The single-sideband carrier being unable to create the envelope of information, which filters the lower side of the signal
    c. Power lines carrying a similar frequency, which interferes with information transfer, as the distance increases from transmission
10. Frequency modulation (FM):
    a. The information signal causes the carrier signal to increase or decrease its amplitude.
    b. The information signal causes the carrier signal to increase or decrease its frequency based on the waveform of the signal.
    c. The information signal causes a power increase on the carrier signal based on the waveform of the signal.

Short Essay Questions

1. What is the name of the branch of the executive government in the United States that regulates frequency allocation and usage?
2. The frequency spectrum in the United States can be broken into three categories. What are they? Provide an example of each.

3. Provide two ways that information throughput can be increased.
4. Explain how amplitude modulation works to deliver a radio signal over a longer distance.
5. Why is AM radio frequency more susceptible to noise or interference than FM?
6. Explain the relationship between frequency and wavelength.
7. Describe the difference between AM and FM modulation techniques.
8. Quadrature amplitude modulation (QAM) is a combination of what two modulation schemes?
9. Define *carrier wave.*
10. Which modulation scheme transmits the difference (or "delta") between two contiguous sampled points in a signal?

Chapter 4

Signaling Formats, Multiplexing, and Digital Transmissions

The conversion of analog signals to a digital format, as discussed in Chapter 3, gave the telecommunications world a new delivery mechanism that opened the door to new and far-reaching technologies. These new technologies could not have been imagined when 1s and 0s started pulsing down copper wires. From cellular data and telephony applications to packets of data moving across the Internet, all have their roots in digital signaling applications.

Understanding the process of conversion previously discussed helps one understand the process of delivering the data from one location to another. The term *data* is used in a broad sense because all voice conversations through core networks as well as traditional data transmissions are represented by the same type of symbol format: 1s and 0s. This chapter will discuss the various formats used to send binary signals along an information path and connect to common digital circuits used in commercial applications today.

Digital Formats

There are a number of different methods of transmitting data through a medium by representing them as electrical pulses. Exhibit 4.1 shows the primary formats used in the communications industry. The most popular, bipolar return-to-zero (RZ), or

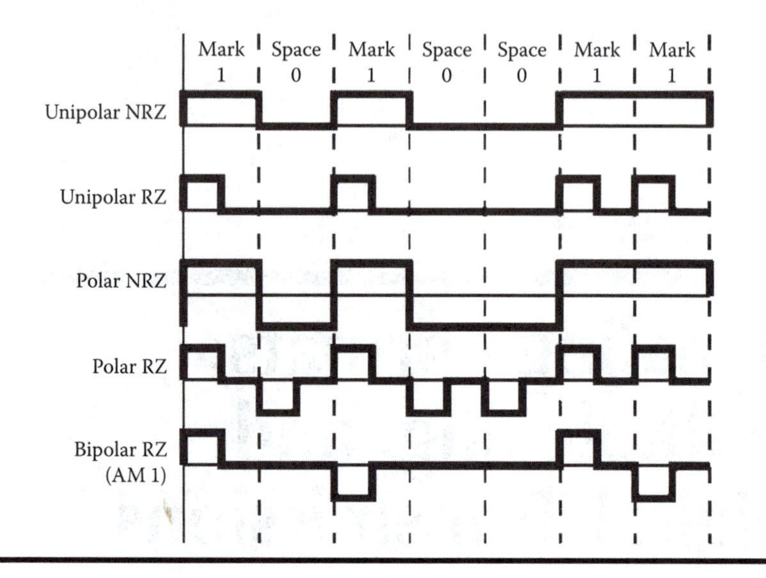

Exhibit 4.1 Binary signaling formats.

alternate mark inversion (AMI), and its relationship to digital information transfer will be discussed at length.

There are two primary ways of delivering this information over a selected medium: half duplex and full duplex. Half duplex refers to the ability of a communications circuit to transmit information in both directions (sender and receiver); however, only one direction of information flow can be accommodated at one time. This requires a system that will help the receiver acknowledge information transfer and delivery, and act as the gatekeeper that allows the return information flow from receiver to sender. Early data transmission circuits relied on this format to allow the passage of information and the confirmation of received data over unreliable noise circuits.

Full duplex allows the transmission of information simultaneously in both directions from sender and receiver. Most data connections today rely on bidirectional data flow to handle the increased throughput demands. The basic local area network (LAN) technology, Ethernet, has expanded its technical offerings over the years to include full-duplex transmission as the demand for higher data throughput increases.

These forms of transmission come in one of the binary formats shown in Exhibit 4.1. Unipolar non-return-to-zero (NRZ) defines its unipolar signaling by keeping the voltage polarity (positive or negative) the same throughout the transmission. The sending and receiving equipment is designed to recognize and detect only the signal polarity defined by the hardware configuration. The NRZ aspect of the signal indicates that the mark, or the indication of a binary 1, is held high, or "on," for a specific time period during the transmission. The time period for which the mark is held on is dictated by the technology being used. The signal does not return to the neutral or nonvoltage condition unless a bit defined as a 0 is transmitted.

Unipolar RZ is an improvement over NRZ because it requires less power to drive the signal, the required mark on time is reduced, and it sets the idle condition of the line signal in a nonvoltage condition. Polar NRZ uses both positive and negative voltage indications to reference the binary signals. Any negative voltage condition can be viewed as a 0 data bit, and any positive voltages can be viewed as a 1 data bit. The progression to polar RZ is evident in Exhibit 4.1.

The final mutation of the information transmission methods is the bipolar RZ. This method changes the voltage indication for every mark that occurs in the transmission. This process is used because of power requirements necessary for transmission and the ability to detect errors in the signal stream more effectively. If two marks appear consecutively in the data stream as positive (or negative) marks, there has been an error in transmission. One of the keys to digital transmission is the synchronization of the data bit stream from sender to receiver in a circuit-switched environment. Recognizing this fact, the transmission equipment looks to the signal stream for timing information. If no information is being transmitted, which translates into 0s or no voltage pulsing down the circuit, timing problems may occur. The most popular ways currently used to notify the receiving equipment that the transmitting equipment is "still alive" is by using either the zero code substitution (ZCS) method or the bipolar 8-zero substitution (B8ZS) method.

ZCS simply inverts the least significant bit in a stream of 0s to indicate a mark. Exhibit 4.2 shows how the before and after condition is represented in a transmission. ZCS goes by a number of other acronyms and names: AMI is used interchangeably with ZCS to define the zero code replacement process. Another term used is Bit7, which also refers to the eighth bit being changed from a 0 to a 1 to keep the signal active. In voice traffic, the insignificance of this data bit being given an improper value has little effect on the conversation quality of the signal.

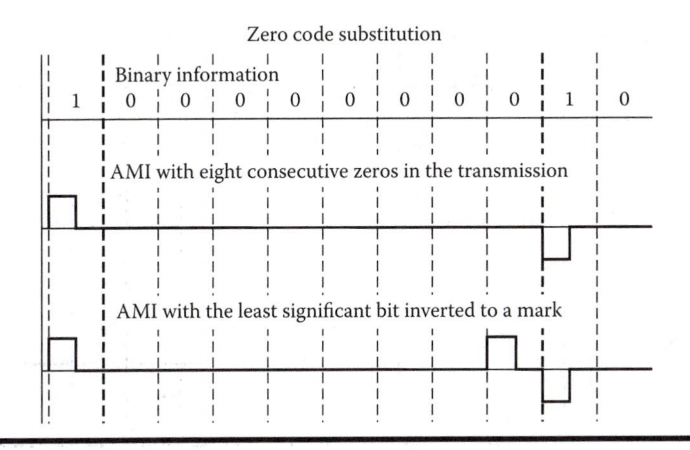

Exhibit 4.2 Alternate mark inversion on a data transmission.

B8ZS uses a different technique to solve the same problem. When it recognizes the same string of 0s, it substitutes a known violation of bits in place of the 0s, which is detected on the receiving end, recognized as a violation, and discarded as a replacement package for the 0 string. Exhibit 4.3 is an example of how B8ZS behaves in a signal transmission; note how B8ZS violates the polarity of the next mark that is inserted into the string of consecutive 0s at slot 4. It follows with the first half of a dual violation in the fifth data slot and another violation inserted into the seventh slot. The eighth slot is also marked and is in correct bipolar indication to the next mark. As previously mentioned, this is the preferred format for keep-alive signaling on digital circuit transmissions. This function is achieved with hardware connecting the transmitting equipment to the wide area network (WAN).

The channel service unit (CSU), or the digital service unit (DSU), is a device used to interface between transmitting equipment and the external circuit in the WAN that carries the information. These devices were once part of the network owned by the telephone companies; however, since divestiture, they have become the responsibility of the companies using the circuits. The CSU/DSU device is used to convert the signal from the transmitting equipment from a unipolar condition to a bipolar 1, and to detect incoming violations and give alarm indications of the errors. The CSU/DSU is also used for loopback testing on the circuit. The device can recognize a predefined bit stream and loop back transmitted data to the sender.

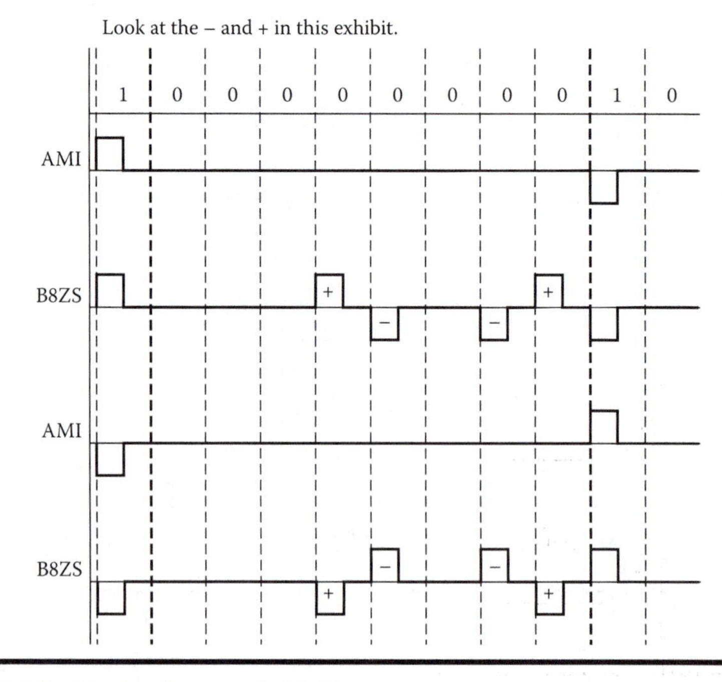

Exhibit 4.3 Bipolar 8-zero substitution.

This process is used to check the integrity of the circuit and to troubleshoot areas of failure on the connection. Knowing how to loop back a data circuit to the CSU/DSU and having established benchmarks regarding its normal operating condition is critical to maintaining a quality data connection.

This information is important because all connections to the WAN rely on the proper formatting and correction process. If a digital circuit is ordered with ZCS keep-alive signaling while the carrier provides B8ZS, then alarms, faulty operation, and flaky problems start appearing in relation to the circuit. If the system is looking for ZCS and a packet of 1s and 0s comes down the line as B8ZS, the system tries to analyze the data bits as information. Because there is an intentional bipolar violation in B8ZS, the receiving system signals alarms because it is not expecting to see those violations in the data stream. Obviously, coordinating and double-checking all network and system programming is critical to network management.

Multiplexing

The two primary forms of multiplexing currently used across both voice and data networks are frequency-division multiplexing (FDM) and time-division multiplexing (TDM). These two methods have been employed to maximize the throughput on a given circuit with respect to frequency for information transmission.

Frequency-Division Multiplexing

FDM was originally developed to provide a way to deliver more voice circuits over a limited amount of copper wires carrying the circuits across the network. FDM divided the available bandwidth into 3000 Hz subchannels to carry voice conversations. These channels were bounded by 500 Hz guard channels and frequency spacing on the low and high ends of the subchannel, and brought together in a multiplexer to be delivered over the common medium. (A multiplexer is a device that combines more than one frequency or circuit in a time- or bandwidth-specific manner to be delivered over a shared medium.) In FDM, the separation is based on the frequencies used for the channel transmission.

FDM has had a number of important uses, ranging from increasing the capacity of voice trunks in the core network to providing the format for microwave transmission applications. However, FDM has been replaced in most applications by TDM because FDM has a number of drawbacks. The primary drawback occurs when a signal is regenerated for extended transmission: noise is amplified along with the original information. Quality issues come into play as the signal is amplified too many times over longer distances, thus increasing the amount of noise sent with the original amplified signal.

Nevertheless, FDM is applicable today with improvements in solid-state technologies used in the amplification process. A digital subscriber line (DSL) is based on an FDM application over a twisted-pair copper cable running to the residential environment. DSL is a high-speed network access technology that provides T-1 rates over an existing copper infrastructure.

Time-Division Multiplexing

TDM is currently the most common form of multiplexing used in North America. Similar to FDM, it combines a number of circuits together to be delivered over a single transport medium. However, unlike FDM and its use of frequency for the allocated separation, TDM uses specifically allocated time slots to deliver the information.

The synchronization of the TDM signal is critical to the delivery of information. Each time slot requires special framing and coordination bits that tell the receiving end where each information slot starts and stops. TDM is based on digital signaling from end to end. Each channel is allocated a specific time slot to deliver information to the receiving end. If no information is available for transfer, no data (0s) are entered. This can be a waste of available throughput capability, a problem that has been solved with a more complex method of TDM called statistical time-division multiplexing (STDM), which uses all available time slots to send significant amounts of information and handles inbound data on a first-come, first-served basis. TDM, however, continues to be the primary method of delivering information over digital circuits.

Digital Circuits

The most widely used digital circuit in North America is currently T-1. Variations of this circuit, digital signal 0 (DS-0) and fractional T-1, are the cornerstones of the digital hierarchy of the communications network. The fiber-optic core network infrastructure is designed to load and transfer the T-1 circuit across the network.

The basic building block of all digital circuits is DS-0, which is equivalent to one 64 kbps circuit used for either voice or data communications. This discussion looks at the process associated with voice signaling, because it is more complicated and the transfer of knowledge to the data communication applications is easier to accomplish. In chapter 3, we discussed Nyquist's theorem, which states that for fidelity, samples should be taken at twice the highest frequency. If a voice circuit carries the bulk of information at a bandwidth of 4000 Hz, the voice circuit would be sampled at a rate of 8000 times each second. Each one of the 8000 samples is represented by a binary number 8 bits (1 byte) in length. If we multiply the 8000 samples by 8 bits, we get the number of bits per second—64,000—transferred in a single DS-0.

DS-0 is the building block of DS-1 or T-1 (these terms can be used interchangeably). DS-1 consists of 24 DS-0s. The throughput of the circuit is based on multiplying 64 kbps by 24, resulting in 1.536 Mbps. The actual throughput of a DS-1 is 1.544 Mbps because there are 8000 framing bits added to the data stream to separate each segment appropriately.

Frequency-Hopping Spread Spectrum

The frequency-hopping spread spectrum (FHSS) is a method in which information is moved from one frequency to another through an 83 MHz width of spectrum (Exhibit 4.4). Both the sender and the receiver must know the code for the hop sequence of the transmitted information. The information is incorporated into a data signal that is broadcast. This broadcast is put into a pseudorandom sequence or hopping code where the signal moves from frequency to frequency. The amount of time spent at each frequency stage is known as the *dwell time*. The dwell time limit is set at less than 10 ms to minimize intercept opportunities. When the signal hops from one frequency to another, the amount of time taken for this is known as the *hop time*. Federal Communications Commission (FCC) standards require that this method use more than 75 frequencies that are hopped and less than 400 ms in dwell time at each of these frequencies.

The bandwidth of the input signal determines the width of each channel that is used in transmission. The bandwidth can be found by multiplying the bandwidth of each hop channel by the number of frequency slots available. The signal is transmitted one portion of an information signal at a time by going through the frequencies as determined by the pseudorandom sequence. When discussing 802.11 wireless local area network (WLAN) transmission that uses FHSS, the hop time uses 300 ms intervals so that you can get a sense of what is regularly used.

There are a few benefits that must be discussed when determining whether or not to use FHSS. It is hard to jam a signal that uses the FHSS method because when someone attempts to jam a signal on one frequency, they will only succeed in knocking out a few bits of information. For the person to succeed in jamming

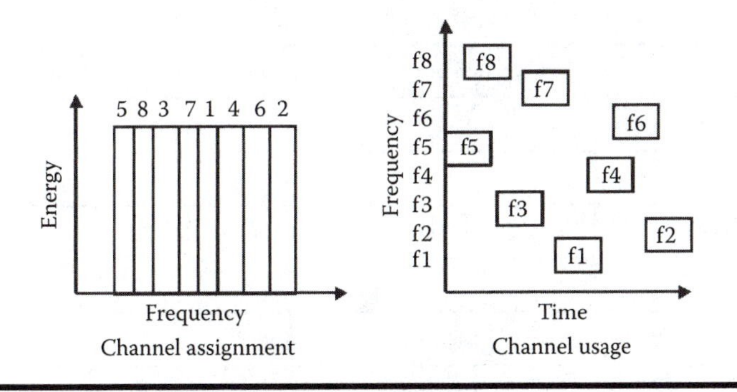

Exhibit 4.4 Frequency-hopping spread spectrum.

FHSS, he or she has to jam all frequencies corresponding to the frequency code. If somebody eavesdrops when the method is used, he or she will only hear unintelligible blips due to the short dwell time on each frequency. Lastly, fixed power reduces the jamming power in any one-frequency band.

Direct-Sequence Spread Spectrum

Direct-sequence spread spectrum (DSSS) is a method used in both 802.11 WLAN processes and code-division multiple-access transmissions associated with mobile wireless networks (Exhibit 4.5). A spreading code is employed to expand each transmitted bit of information into a larger number of bits. The success of the method is directly related to the chipping rate of the spreading code. In other words, as the ratio of chipping rate to information rate increases, it is less susceptible to interference. This expanded-bit section is referred to as the *chip set* or *chips*. A 2.4 GHz spectrum with 11 different frequencies centered within that spectrum is available. Each of these frequencies is 22 MHz wide, and they cannot overlap with one another. When using WLAN processes, the frequency channels used are 1, 6, and 11, where channels are five frequencies apart to eliminate interference. Once a channel is chosen, transmission continues on that channel until it is completed. The advantage of this is that more data can be sent by using a single transmission space; however,

	Combination of data and chipping code at the transmitter							
Information bit	0							
	↓							
Chipping code	0	1	1	1	0	0	1	0
	↓							
Transmitted data	1	0	0	0	1	1	0	1
	Combination of received signals and chipping code at the receiver							
Received data	1	0	0	0	1	1	0	1
	↓							
Chipping code	0	1	1	1	0	0	1	0
	↓							
Information bit	0							

Exhibit 4.5 Direct-sequence spread spectrum.

there is a greater likelihood of interference, unlike in direct hopping. The chipping codes for the bits of information vary in size but, in a WLAN, are designed so that even if half the information is lost, enough information remains to replicate the original information sent. This is accomplished by having the received signal and chipping codes undergo a test (mathematical computation) that would provide an evaluation on the data transmitted. If the binary information "looks" like a 1, then it will be assumed that the piece of data is a 1.

Summary

Modulation schemes of various complexities provide for the full utilization of available bandwidth. Because most applications today are being quickly adapted to wireless delivery methods, using the available spectrum is critical. However, with every new modulation scheme, the fundamental theories of Shannon still apply. Newer methods are also built upon what we already have in place, such as amplitude, frequency, and phase modulation.

Review Questions

Multiple Choice

1. Which of the following is used for error detection?
   a. B8ZS
   b. AMI
   c. ZCS
   d. All of the above
2. Which of the following throughputs is the rate of the actual data in a T-1 circuit?
   a. 1.544 Mbps
   b. 1.864 Mbps
   c. 1.544 kbps
   d. 1.536 Mbps
3. Which binary format is the most efficient and most widely used?
   a. Unipolar NRZ
   b. Bipolar RZ
   c. Polar NRZ
   d. None of the above
4. What are CSU/DSU devices not used for?
   a. Detection of incoming violations
   b. Determining the number of trunks needed for information transmission
   c. Interfacing the transmitting equipment and the external circuit
   d. Carrying information in a WAN environment

5. Frequency-division multiplexing (FDM) has the following characteristics:
   a. Divides bandwidth by frequency to send more information
   b. Sends more information than TDM
   c. Labels the individual packets with a time stamp
   d. All of the above
6. Time-division multiplexing (TDM) has the following characteristics:
   a. Wastes a channel if that particular channel has no information to send
   b. Sends more information than FDM
   c. Is the most common form of multiplexing in North America
   d. All of the above
7. If the highest frequency in a voice circuit is 3500 Hz, how many samples will there be?
   a. 5000
   b. 6000
   c. 7000
   d. 8000
8. If a DS-0 has a throughput of 64 kbps, what is the total throughput of a single phone line?
   a. 56 kbps
   b. 80 kbps
   c. 64 kbps
   d. 128 kbps
9. If a DS-0 has a throughput of 64 kbps, what is the actual throughput of a single phone line? (Hint: similar to the modem.)
   a. 56 kbps
   b. 80 kbps
   c. 64 kbps
   d. 128 kbps
10. Which bit stream would best represent B8ZS?
    a. 0 0 0 0 0 0 0 –
    b. 0 0 0 + – 0 + –
    c. 0 0 0 + – 0 – +
    d. + 0 0 0 0 0 0 0

Matching Questions

Match the acronym or value on the left with the statement that best describes it on the right.

1. ZCS	a. Actual throughput of a T-1
2. TDM	b. Equivalent of a T-1 over the PSTN
3. CSU/DSU	c. Equivalent of a telephone line
4. DS-0	d. Changes the last bit to keep the signal alive
5. 1.544 Mbps	e. For detection of error based on a polar violation
6. AMI	f. Bandwidth of a voice circuit
7. FDM	g. Divides a specified frequency to send more information
8. 1.536 Mbps	h. Ensures transmission across a WAN
9. DSL	i. Total throughput of a DS-1
10. 4000 Hz	j. Uses specifically allocated time slots to deliver information

Short Essay Questions

1. Explain the differences between half and full duplex, and why one transmission method might be chosen over another.
2. Excluding AMI, what would be some possible advantages and disadvantages of the remaining four binary formats that are all seemingly doing the same thing?
3. Why has bipolar RZ (AMI) been designated the most widely used binary signaling format?
4. What do ZCS and B8ZS have in common, and how do these signals work?
5. What, if anything, is the difference between AMI, ZCS, and B8ZS?
6. What is the correlation between CSU/DSU and AMI, ZCS, and B8ZS?
7. What is a major problem with a digital circuit using ZCS as a keep-alive signal and that signal's carrier using B8ZS signaling?
8. What are the differences between FDM and TDM; which multiplexing method would be more effective, and why?
9. What might be a major disadvantage of TDM?
10. Explain how we arrive at 1.544 Mbps as the throughput for a T-1 line.

Chapter 5

Legacy to Current-Day Telephone Networks

In this chapter, we will explore the evolution of telephone networks from the public switched telephone network (PSTN) to the present-day use of Voice over Internet Protocol (VoIP). The PSTN has provided over 100 years of reliable telephone service to the North American population. Divestiture of the Bell System caused various technology markets to explode, expand, contract, consolidate, and reemerge within the industry, sparking innovation for years. The business telephone market provided technology advances that have set the pace for the competitive nature of the industry. But the PSTN is quickly becoming a legacy technology. Future systems are converging with data networks where the connection to the wide area is dependent upon the convergence of both voice and data services (e.g., Voice over Long-Term Evolution [VoLTE]).

As the reliability and consistency of the legacy telephone network have started to drop with age and technology has advanced, new and more efficient ways of voice conversation are making their way into the picture. This is happening through the deployment of VoIP, a packet-switched technology that combines voice, data, and multimedia communications over an Internet Protocol (IP) environment. As the popularity of VoIP grows, it becomes more of a reality that it will replace the circuit-switched analog PSTN in the very near future. It is important to understand the fundamental differences in these technologies, how they work, and what they can offer to users as we migrate to an all-IP network.

Circuit Switching vs. Packet Switching

One of the most drastic changes in the move toward VoIP is the change from circuit switching, the basis of transmission in the PSTN environment, to packet switching. Circuit switching sets up a single dedicated connection, or circuit, between end points through which users exchange a communication. In packet switching, a communication is broken down into small blocks of data called packets, which are sent along various paths to the destination. The packets are reassembled at the receiving end in the correct order.

Legacy of the Circuit-Switching Network

Circuit switching establishes a dedicated connection between the sending point and the receiving point that is not shared with any other signaled information. This is how the PSTN works. Circuit switching is highly reliable, provides a secure communication link, and is ideal for the distribution of data with a quality of service (QoS) related to the throughput. QoS refers to the capability of a network to provide managed bandwidth and better service to preferential network traffic.

Circuit switching gives any end point the capability to communicate with another end point anywhere in the network through a dedicated connection. This environment, however, does not allow for multiple connections to a single point at the same time for simultaneous information delivery. Exhibit 5.1 illustrates the connections associated with a circuit-switched network.

The setting up and tearing down of connections that are not permanent to the end points are termed switched virtual circuits (SVCs). The process of setting up the communication link between the end points requires the establishment of a connection prior to information transfer. Exhibits 5.2 through 5.5 illustrate the setup process.

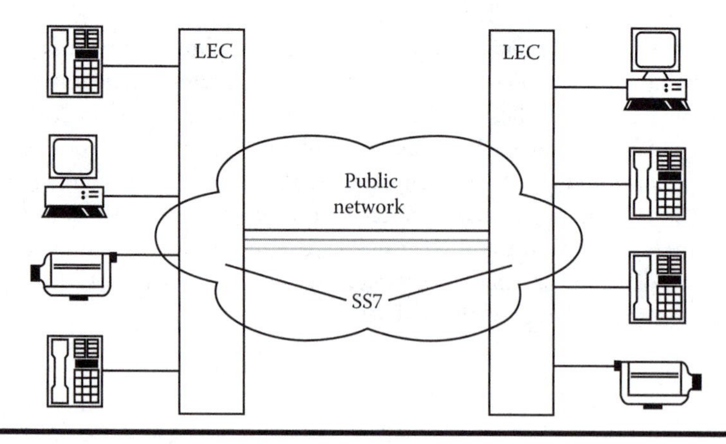

Exhibit 5.1 Circuit-switched network.

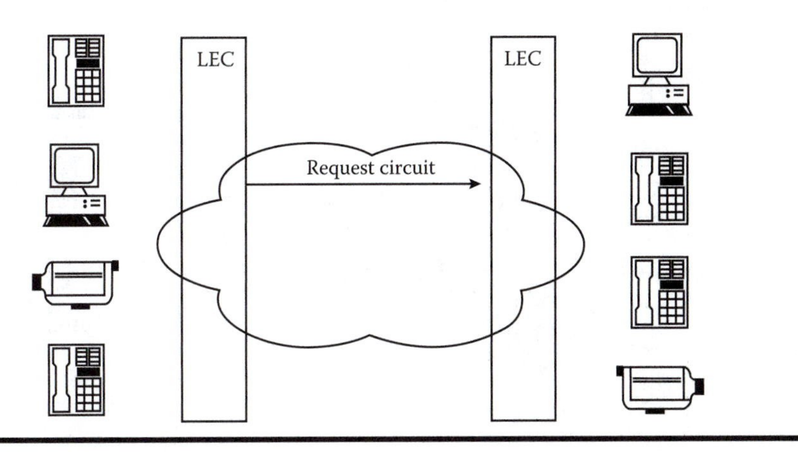

Exhibit 5.2 Circuit switch request.

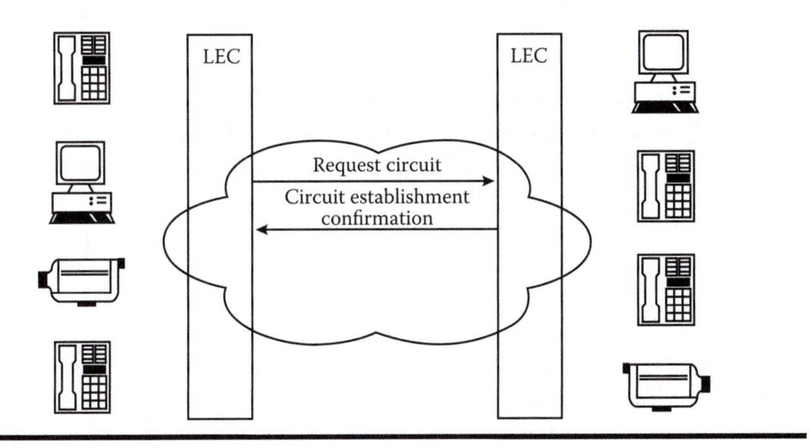

Exhibit 5.3 Circuit switch confirmation.

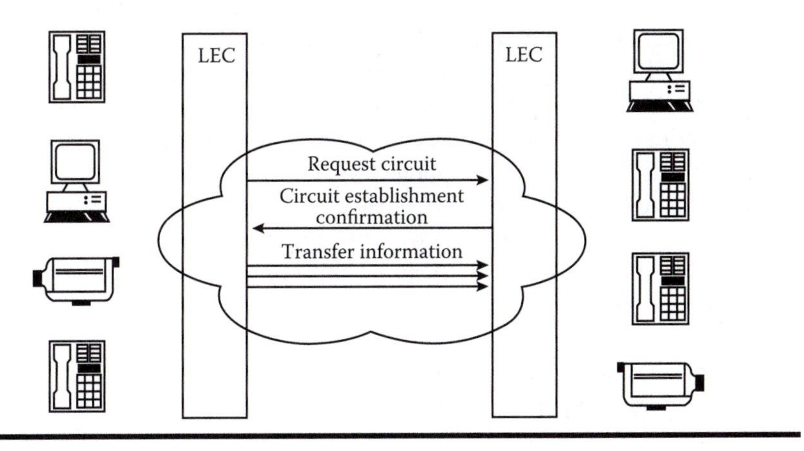

Exhibit 5.4 Circuit switch transfer.

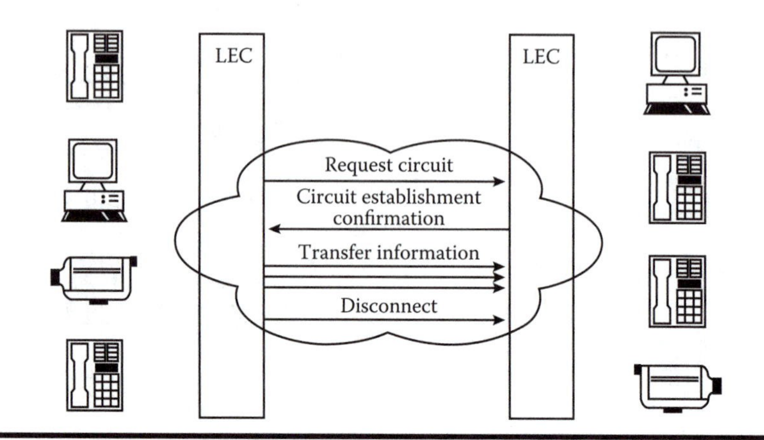

Exhibit 5.5 Circuit switch disconnect.

Circuit switching and the PSTN were originally designed for analog transmissions. New technologies were developed to accommodate digital transmissions. Below is a discussion of Signaling System 7 (SS7) and Integrated Services Digital Network (ISDN). Another digital-capable technology, Asynchronous Transfer Mode (ATM), will be discussed later.

Signaling System 7

SS7 was developed and implemented in the late 1970s and early 1980s to provide the connection management for an all-digital network. The eventual migration to SS7 created an out-of-band signaling method for the PSTN. *Out of band* refers to the fact that all the information for circuit acquisition, setup, and teardown exists in a frequency band (in this example, a network) different from the path taken by the transmitted signal. Before SS7, all the information about the circuit connection was passed in band or within the band of frequency in which the information was being sent. This posed a problem in that the information to direct and charge for toll calls could be fraudulently injected into a signal. This faked signal allowed the caller to bypass long-distance telephone charges.

SS7 provides supervision, address signaling, call progress information, and alerting notification across the network. Its only function is to provide signaling between switches on the PSTN. There are three primary points affiliated with the SS7 network:

1. Service switching points (SSP)
2. Service transfer points (STPs)
3. Service control points (SCPs)

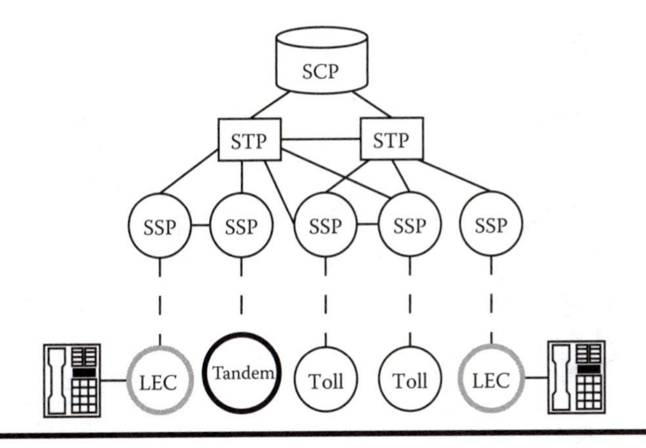

Exhibit 5.6 SS7 block diagram.

SSPs are the locations closest to the user. They are points at the central office (CO) or tandem switch on the PSTN at which the initial information about a call is collected and passed along the network. In some cases with newer technologies (e.g., Interactive Intelligence call center switch), this switching point is directly accessible from the customer premise equipment (CPE).

STPs can be considered points on the SS7 network that transfer link information to the next switch to help establish the circuit. STPs are established in a redundant configuration to ensure the reliability of the network. If an STP fails, at least one redundant STP steps in and performs the connection management in place of the failed STP. The STP gets its guidance information about the call from the switch control points.

SCPs are databases containing all the addressing information about a particular area of the country. These databases can be maintained by either the telephone companies or private enterprises providing services to the PSTN. SCPs provide the information that makes emergency 911 calling, caller identification, and other network-related features possible. The most important idea to remember about SS7 is that it is an out-of-band, stand-alone network used to set up, monitor, and tear down switched circuits on another network. Exhibit 5.6 illustrates a simple block diagram of SS7. Note that each point is connected with more than one link for redundancy purposes.

Integrated Services Digital Network

ISDN is a digital circuit-switched network system that can provide voice, data, and video (multimedia) services over a PSTN connection (Exhibit 5.7). ISDN relies on SS7 to carry its connection information from source to destination in approximately 2 seconds across the North American network. The all-digital nature of

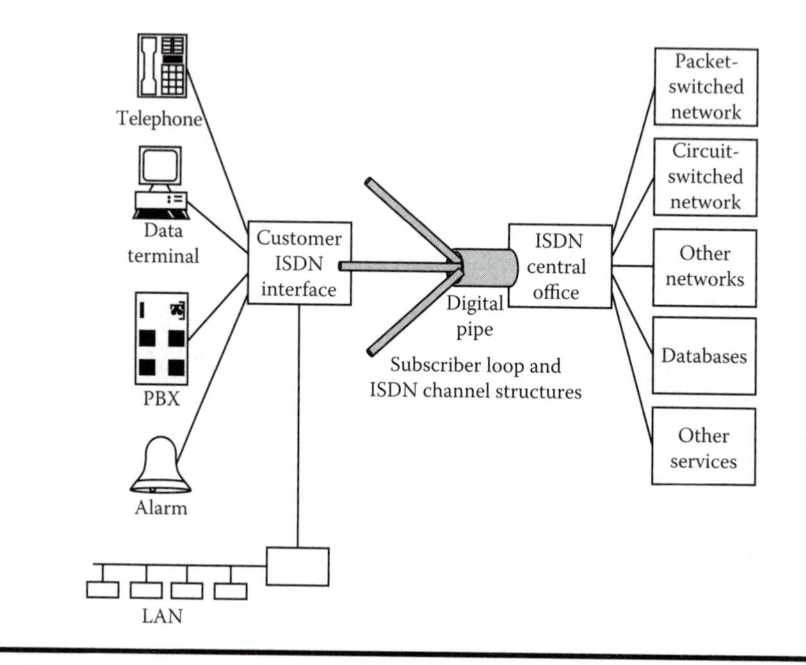

Exhibit 5.7 ISDN applications.

ISDN allows the signal to stay in its digital form without the need to change from analog to digital and then back to analog again. Although other high-speed data connections have surpassed ISDN in throughput capacities, it continues to be used as an excellent backup direct-dial connection for data networks because it is part of the PSTN.

There are two types of ISDN predominantly in use today: Basic Rate Interface (BRI) and Primary Rate Interface (PRI). BRI consists of two-bearer (B) channels of 64 kbps throughput. The two channels are managed by one 16 kbps data (D) channel. The standard throughput for a BRI ISDN circuit is that of the two B channels, or 128 kbps. Bonding, also called *H-channels*, is the use of multiple BRI circuits together to provide greater throughput.

PRI is a 24-channel, 1.544 Mbps circuit in North America used in high-throughput applications. The B-channel capacity is the same as on the BRI circuit, with 64 kbps for each channel, and the D channel's capacity is 64 kbps. Of the PRI circuit channels, 23 are used for the defined application, with the 24th channel reserved as the D channel to control framing, signaling, timing, supervision, and other circuit control issues for the other 23 channels.

The switched nature of PRI has helped eliminate the need for point-to-point T-1 circuits in many networks. The PRI connection is treated as a long-distance telephone call for connections made out of the source's local access and transport area (LATA). PRI circuits can be designed to have portions of the service used for

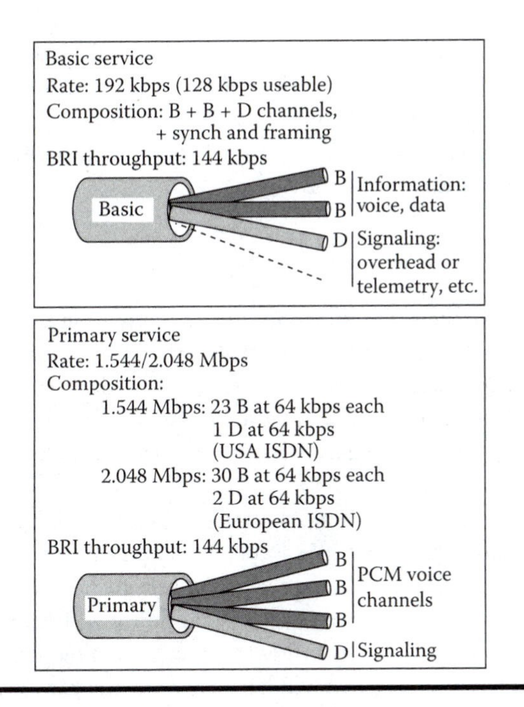

Exhibit 5.8 ISDN configurations.

the data network, some for the voice needs, and still other channels defined for video conferencing services. The channels can be configured to be used on demand by a system administrator, who may need more data throughput or video services for a specific application. Exhibit 5.8 gives a representation of the two ISDN channel configurations.

Circuit switching, while reliable for streaming communications, is not optimal for digital and Internet traffic. Dedicated connections can be inefficient, and increases in capacity require expensive equipment upgrades. For such reasons, circuit switching is being eclipsed by the packet-switching method.

Packet Switching

Packet switching is different from circuit switching in two distinct ways. First, packet switching breaks the communication into chunks (packets) of data to be sent, instead of a sending one constant stream of data. Second, packet switching does not require a dedicated circuit for communications. It sends data when needed and can share the path with traffic from separate user transmissions, making it a much more efficient use of bandwidth than circuit switching.

Packet switching is great for "bursty" data, like e-mail or web pages. When a web page is requested, packets are sent back and forth between a client and web server to load the web page. Once the page is loaded, the data transfer is complete.

This is much different from transmitting streaming data, like a voice call, using packet switching. There is a constant flow of data being transmitted and received, which makes using packet switching more difficult but still possible. Below is a description of how packet switching works, which is an introduction to how VoIP works on a packet-switched network.

Packet Headers

Packetized data, or packets, are able to find their destination by the use of a packet header. These headers are read and forwarded to their destination by devices in the network and key connection points called *routers* (or nodes). The header of a packet contains information such as the source (or sender's) address, the destination (or recipient's) address, how large the packet is, the sequence number of the packet, and error-checking information.

The source and destination addresses are used by the packet to tell nodes where the packet came from and where it is going. They work much like an address and return address on an envelope. A router tries its best to forward the packet to the destination address, but if, for some reason, it cannot, it will send an error message back to the source. When a packet is created, it is given a sequenced identification number so that the receiving client will be able to put the packets back in order or tell if one is missing. The first packet created is usually number 1, the second is 2, and this repeats for as long as a single application is sending data. Different types of connections use the sequence numbers in different ways.

Quality of Service

QoS refers to the capability of a network to provide managed bandwidth and better service to preferential network traffic. These services include bandwidth, jitter, delay, packet loss, and desequencing. When applied to a packet-switched networking, QoS indicates the ability to guarantee an accepted level of packet loss. It is impossible to predict the paths taken by packets, so it is likely that some sort of incident will occur in the packet's transmission. QoS helps differentiate network flows and reserve a portion of the bandwidth for those requiring continuous service without breaks.

Service level essentially means the required levels of network capacity. In QoS, there are three levels defined: best effort, differentiated service, and guaranteed service. Best effort is the lowest level and shows no differentiation between network flows or guarantees. Differentiated service defines priority levels for different network flows with a possible guarantee. Allocated priority to voice ensures that delay-sensitive applications are not compromised by other applications and also ensures that voice packets will be delivered without loss. Guaranteed service reserves network resources for certain flow types. To reserve these resources, the resource reservation protocol (RSVP) is used. RSVP is a signaling protocol within the Transmission Control Protocol (TCP)/IP suite of protocols used to manage data packet transfer on the network.

Connectionless vs. Connection Oriented

Packet switching can transmit packets using one of two methods: connectionless, or unreliable delivery, and connection oriented, or reliable delivery. Connectionless transmission is called unreliable because there is no acknowledgment between the sender and receiver when packets are transmitted. It is a best-effort transmission. The sender sends his/her packets off through the network and hopes that the receiver receives all of them. While this method can be more unreliable than the connection-oriented approach, it does, however, increase the speeds of data transmission by cutting down on overhead, or management data, for the packet delivery process.

Connection-oriented transmissions are a reliable form of packet transmission because there is a constant acknowledgment process that takes place between the sender and receiver. One of the most used connection-oriented protocols on the Internet is TCP. To transmit packets using TCP, the sender and receiver must complete a "three-way handshake," where they exchange three messages confirming that each party is accurately receiving them. Then the sender is cleared to transmit data. The receiver checks the sequence number of each packet to make sure that he/she is receiving each packet in order. If a packet is not received, or received out of order, the missing packets are resent, starting from the packet that caused the error. It is a lengthy process that can take much longer than the connectionless approach, but it ensures that there are no errors during packet transmissions. You will learn more about packet switching and TCP in Chapters 7 and 8.

Asynchronous Transfer Mode

ATM is a compromise between circuit switching and packet switching. It is a cell-based, asynchronous delivery technology designed to carry voice, data, and multimedia services over the same link simultaneously. It can emulate a traditional voice network and provide synchronous delivery of a voice conversation, or it can be used to deliver bursty asynchronous data traffic over the wide area. ATM is capable of providing connection-oriented services (e.g., voice) and connectionless services (e.g., local area network [LAN] emulation) over the same network.

ATM Architecture

Like circuit-switched systems, ATM establishes a connection between the source and destination, similar to a traditional telephone voice call, before it transmits information. However, by defining its throughput on specific cell sizes, ATM resembles a packet network (e.g., Ethernet). ATM cells define their own route through a network, but they are defined by a connection-oriented condition.

ATM is the primary choice of technology for a number of applications when QoS is required for the data throughput. It can be used for workstation-to-workstation computer-aided design (CAD) and computer-aided manufacturing (CAM), in which large files of graphical information must be transferred from one engineering site to another. ATM is used in the backbone of campus networks to deliver multimedia to the desktop and to handle heavy client/server traffic. It can be implemented in the metropolitan area network (MAN) to connect different locations of the same enterprise with high-speed, high-bandwidth, and high-quality service. It is also implemented in the carrier network backbone for reliable switching performance on the long-distance network.

Although emerging technologies have been touted as replacing ATM in the network (e.g., Gigabit Ethernet), ATM has QoS that cannot be matched by other technologies; its capability to deliver different services over the same circuit is unique.

Legacy Public Switched Telephone Network

Understanding the basic delivery mechanism of the PSTN (the hierarchical configuration of the switching offices), the CPE that connects to the network, and the basis of circuit-switched technology allows you to become fluent in a language spoken, until a few decades ago in this country, by only the members of the Bell family.

The PSTN has been a part of our daily existence for over 100 years. The historical significance of Alexander Graham Bell's invention, or possibly his ability to patent his idea only moments before his competitor Elisha Gray, has been the cornerstone of communications technology from the late 1800s through the 1990s. The PSTN is based on the creation of a direct connection from the sender to the receiver over an extremely sophisticated delivery network comprised of copper, fiber optics, satellites, and fixed wireless and mobile wireless circuits. The network is comprised of five basic components (with numerous subsets): the telephone, network access, COs, trunks and special circuits, and CPE. The purpose of this section is to provide a firm understanding of the basic components of the PSTN.

The Telephone

The local loop or subscriber loop is the equipment used to connect residential and commercial service to the telephone company's CO. This connection consists primarily of copper twisted-pair cabling. Exhibit 5.9 shows a block diagram of the subscriber loop.

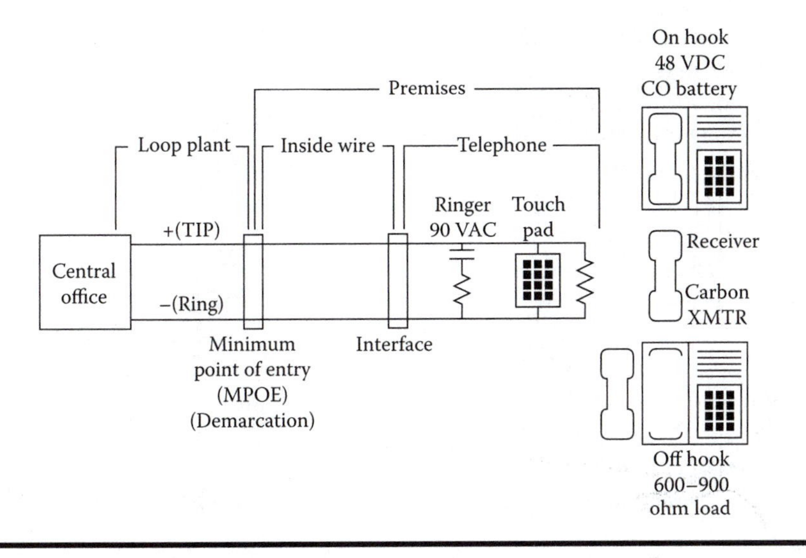

Exhibit 5.9 Subscriber loop.

The telephone that we have become accustomed to as an everyday commodity of service acts much the same as a light switch. To toggle the dial tone "on," the switch is thrown by lifting the handset, which is commonly referred to as going off hook. Within the inner workings of the telephone, a switch is closed that provides current from the CO to the electronic components of the telephone. The circuit consists of two wires or a pair of copper wires ranging from 18 to 24 AWG (American wire gauge). When the tip side of the line measured with a voltmeter to an earth ground reads zero voltage, the ring side of the cable pair reads –48 V direct current (DC). The -48 V DC connection provided by the copper wires strung to your location from the telephone company is delivered with negative voltage because it provides for greater resistance in the ionization of the materials in the copper cable. This helps reduce the corrosion of the wire. The two terms, *tip* and *ring*, originated from the component parts on the plugs used by the early telephone operators who connected one party to another.

The transmitter acts like a microphone, converting analog wave signals that excite a diaphragm, which vibrates carbon chips to create electrical signals that are transmitted over the telephone cable pair. Modern electronic subsets (telephones) use a similar method of generating a signal with more sophisticated electronic components. The receiver's responsibility is to replicate the transmitted electrical vibrations back into a decipherable, audible signal. Legacy networks primarily started in an analog environment. Modern digital equipment converts the transmission from the analog signal at the CO to a digital signal that is sent over the network to the receiving party's CO. Finally, the signal is returned to an analog format and is sent to the receiver's subset. This "last mile," or the connection from the CO to the end user, is the bottleneck for the delivery of high-speed services. The copper environment was engineered to provide reliable service for voice-grade transmissions within a 4000 Hz bandwidth. Various components and configurations (e.g., coils and taps) caused considerable problems in trying to deliver high-speed data connections to customers.

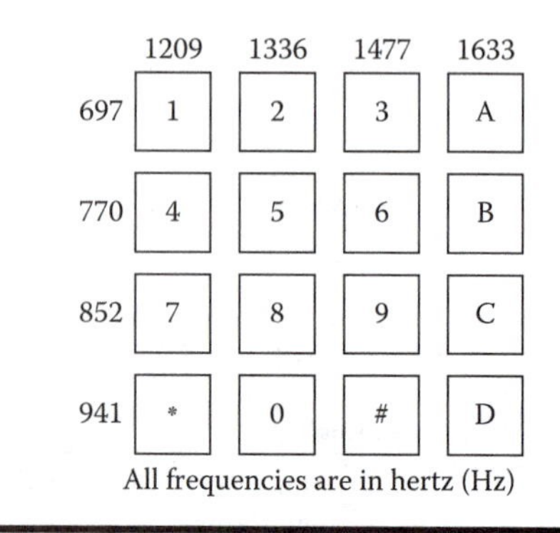

Exhibit 5.10 Dual-tone multifrequency dial pad.

The signaling to the CO from the subset could be based on two methods: dual-tone multifrequency (DTMF) and rotary (or pulse) dialing. DTMF or touch-tone provides two distinct frequencies for each key that is pressed on the dial pad. These frequencies are collected at the CO and used to define the called party. Exhibit 5.10 shows the different frequencies and their alphanumeric association.

The second format (which has all but disappeared from the network in North America) is the rotary- or pulse-dialing signal. Rotary dials relied on a make/break condition of a set of relays on the dial to open and close the circuit to the CO. The ratio of connection to disconnection (make/break) was 60:40. The on/off pulse of the subscriber loop signaled the CO equipment to indicate the number dialed. The pulse period is 100 ms in duration, 1 pulse for each digit and 10 pulses for the digit 0. Exhibit 5.11 graphs the pulse signal and the make/break ratio.

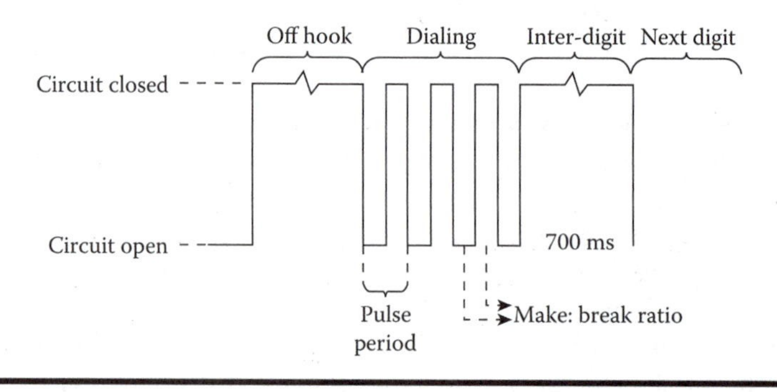

Exhibit 5.11 Pulse dialing.

When the signal is received in either DTMF or pulse configuration, the CO pulls the signal from the circuit and passes it along to an overarching signaling network that coordinates all the setup and teardown of calls in North America: SS7. Recall from the circuit-switching section that SS7 is an out-of-band method of transferring the call setup information sent from the calling party to a receiving station. *Out of band* is defined by the use of a network other than the one in which the voice conversation travels over to secure a circuit.

Network Access

Network access is defined by a number of participants, ranging from what is left of the Regional Bell Operating Companies (RBOCs), other independent local exchange carriers (LECs), interexchange carriers (IXCs), cellular operators, and competitive local exchange carriers (CLECs). Sorting out the acronym soup associated with the service providers is a good starting point in understanding the differences between the players. The LEC is the entity that provides the last mile of service from the CO to a residence or business. The LECs had control over the provisioning of dial tone until the Telecommunications Act of 1996 forced the market open to other CLECs. There are a few companies remaining competing with the LECs to provide dial tone to major metropolitan areas.

Exhibit 5.12 shows the progression from the end office to the switching center that directly services the residential and business circuits, up to the regional offices that connect major portions of North America. Since the advent of digital switching in networks, this hierarchy has become flatter, with the sectional and primary layers bypassed in more advanced network configurations. The toll office also represents a tandem switch that acts as a trunk aggregator between end offices. A tandem

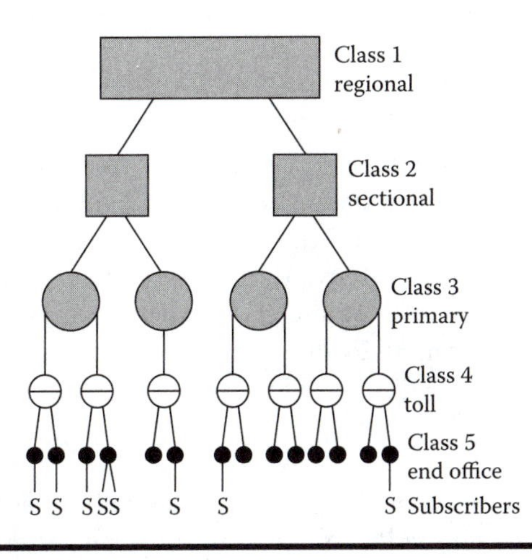

Exhibit 5.12 PSTN tiered offices.

switch does not provide dial tone to an end user; rather, it acts as a switching point between end offices. It is used to eliminate mesh networking between end offices in large metropolitan areas.

With wireless local loop (WLL) or fixed wireless access (FWA), companies bypass the use of LEC infrastructure by using broadband wireless connections between their points of presence (POPs) and customer locations. The technology originally fell into two major categories: local multipoint distribution services (LMDSs) and multichannel multipoint distribution services (MMDSs). LMDS and MMDS license holders were primarily focused on delivering bundled-services voice, data, Internet, and multimedia access to metropolitan customers. The principle of using this technology was sound; however, combinations of equipment malfunctions, antenna deployment, and property management resistance to wholesale services made the penetration necessary for profitability difficult. The higher the speed (the more hertz per second), the more line of sight (LOS) is necessary for delivery of the signal. Weather, physical obstructions, and municipal restrictions on antenna placement slowed the growth of these technologies.

The IXC is also known as a long-distance provider. AT&T, WorldCom, Qwest, Global Crossing, Sprint, and other players completely changed this market segment from a single provider of services to one that provides profitability to numerous companies at a lower cost to consumers. Equal access to providing toll services allowed numerous companies to enter this segment of the business.

Cellular providers are now an integral piece of the network access puzzle. With the advent of digital wireless services for voice and data connections, cellular providers have been able to get early technology adaptors to "cut the cord" with wire-line LEC services. The progression to even more advanced technology with fourth-generation wireless (4G) and fifth-generation wireless (5G) has accelerated the residential users abandoning of wire-line telephony services. A greater examination of these technologies will be covered in the chapter on wireless technology.

Trunks and Lines

The CO's main function is to provide a connection from the end user, residential or business, to the PSTN. There are five main types of available circuits consumers utilized in provisioning services for their personal use or business application. These circuits are loop-start, ground-start, direct-inward-dial, and "ear-and-mouth" (E&M) trunks, and Centrex. These circuits and their analog services are now considered legacy but will remain in use for years to come.

Loop Start

Loop-start lines are the basis of all residential and small- to medium-size business services provided through the PSTN. Loop-start lines have immediate dial tone. The circuit is ready to be dialed out on 24 h a day, 7 days a week, all year long.

When we go off hook with the handset of our telephone, we open the switch to allow the dial tone carrying voltage to be sent to the receiver. Small- to medium-size businesses use these circuits on key systems that present more than one outside line to the user. Benefits of loop-start lines include cost-effectiveness and faster deployment than other types of circuits.

Ground Start

Ground-start trunks are used for private automatic branch exchange (PABX or private branch exchange [PBX]) connections to the CO. PABXs will be discussed in a later section. These circuits are associated with PSTN service for medium- to large-size businesses. The circuit pair does not have the immediate dial tone that is associated with the loop-start trunk. Instead, the CO waits for the PABX to send a signal (ground) to close a relay at the CO that allows the dial tone to be sent to the end user. The purpose of this type of trunk is to keep glare, or collisions, from occurring between outbound callers on PABX and inbound callers to the business. Most business environments rely on dialing a code (usually the digit *9* in North America) from a subset to access an outside line. If the trunk were always hot, as in the case of loop-start trunks, it would allow inbound traffic to collide with outbound access. In most states, the public regulatory commissions dictate the use of ground-start trunks on PABX based on the Federal Communications Commission's (FCC) registration number filed in compliance with ordering circuits from the LEC. Ground-start circuits (and other business-defined circuits) cost more than the residential service.

Direct-Inward-Dial Trunks

Direct-inward-dial (DID) trunks are used with PABX or some of the more advanced hybrid systems available on the PSTN market. These technologies will be discussed later. The purpose of DID trunks is to allow a large block of numbers to be assigned to users on the customer side of PABX from the CO, while connection between the PABX and the CO is much smaller than the numbers assigned. A trunk circuit that gains its power from the PABX (instead of the CO) "looks" at the number that the CO forwards to it. After this, the number is compared to a table that defines the call's internal station destination. The circuit connection ratio to the amount of numbers assigned can range from a 4-to-1 structure down to a 2-to-1 configuration, depending on the level of inbound traffic. These trunks are designed to give external callers direct access to individuals or departments within large corporations.

E&M Trunks

E&M trunks were originally designed to connect multiple PABXs separated in wide area conditions. The E&M circuit is designed to seize dial tone from the remote site and allow the calling party to have a remote station dial tone from the distant PABX to reduce the toll traffic within a corporation by using the always-on E&M circuit. The advent of lower-priced long-distance services and direct-connect T-1 carrier services largely eliminated the demand for E&M circuits. The cost of the E&M circuit was based on wire distance (in miles) between the locations and could be very expensive if the circuit spans a great distance (e.g., state to state). Some carriers continue to define the connection of their digital trunks with E&M parameters because the circuit is a switch-to-switch connection.

Centrex

Centrex is more of an extension of the CO than an actual trunk. Its features act similarly to those of PABX, except that the features reside in the CO switch rather than in hardware located on the customer premises. This type of circuit was originally designed for companies that had scattered locations across a metropolitan area needing a common dialing plan. Centrex callers dial four or five digits to call within their dial plan. For example, a police substation can call another substation or the main jail by dialing a four-digit station number. Dialing out of the network requires an access code (e.g., *9*) for routing outside of the dial plan. Historically, Centrex configured in this manner has been an expensive proposition with limited numbers of lines. In large system configurations, there is a definite advantage to the Centrex service over PABX installation, especially if the company is scattered across a wide geographic area.

Centrex has also been marketed toward small- to medium-size businesses by the LECs under various plans that ensured that the customer stayed "connected" to the LEC. The PABX features of the Centrex service can be advantageous to these businesses: transferring inbound calls back out to another location or a mobile user, conference calling, and station dialing features to an associated business location. Cloud-based voice services fundamentally follow the Centrex model of delivery but based on IP packet-switched standards.

Customer Premise Equipment

CPE is a relatively entrenched portion of the PSTN industry. Various components make up the CPE landscape. Small- to medium-size businesses today use digital key systems to handle multiple inbound and outbound trunks, while the larger corporations rely on PABX to handle the larger traffic demands that their companies generate. Wedged between these two entities is the hybrid system that has the simplicity of the key system and the feature-rich environment of the PABX.

Private Automatic Branch Exchanges

PABX, also known as PBXs, in its simplest configuration can be viewed as a specialized computer that is used to connect inbound voice traffic to the appropriate destination, route outbound calls to available trunks, and allow station-to-station communication. PABX is characterized by feature-rich offerings that process voice calls. They are used to consolidate numerous internal users to a reduced set of trunks connected to the PSTN. Peripheral and system-supported services such as voice processing, station message detailed recording (SMDR), automatic call distribution (ACD), and least-cost routing (LCR) have been created to handle and record call traffic in a highly efficient manner.

PABX equipment can connect to analog as well as digital trunks from the LEC or IXC. The collection of information on how outside trunk traffic is handled, the duration of calls made, the number of calls made, and other important traffic considerations are generated by SMDR. Various software programs are used to pull out information on single-user activity, and corporate-wide traffic can message SMDR output. Busy hour, busy day, busy time of year, and other management staffing considerations, along with monitoring for internal toll abuse, are important aspects of SMDR.

ACD is a system capability that allocates incoming trunk calls among agents in a programmed group in a way that each agent receives an equitable share of the load. ACD is the cornerstone of the call center (or contact center) industry. If you have ever called L.L. Bean or another catalog company, or have dialed into a customer service center, your call has been processed by an ACD system. There are highly specialized stand-alone ACD systems (e.g., Interactive Intelligence, Aspect), which are designed specifically for the call center environment. Many higher quality PABXs include ACD software in their standard software configuration to handle varying levels of inbound and outbound traffic.

The ACD can be configured to work as a subset of the PABX-served users. An example of this is creating a customer service center within a corporation for external clients with the ACD, while the rest of the corporation works from a traditional configuration on the system. There are uses within companies today where call traffic must be processed in a queued fashion in order to efficiently handle large volumes of traffic. For example, short-staffed human resources offices become inundated with calls after a change has occurred with a monthly payroll deduction. Handling the calls in a timely, orderly manner is the hallmark of ACD.

Exhibit 5.13 shows a logic flow of how a call is analyzed and processed within an ACD program. The flow of the call after it rings into the group to which it is routed is depicted by the flow diagram in Exhibit 5.14.

LCR or automatic route selection (ARS) is the automatic selection of the most economically available route for each outgoing trunk call. Route selection can be based on numerous factors, including the time of day, day of week, destination area code, class of service of the caller and trunk, and status of the trunk. LCR can be used to absorb dialed digits, modify the outbound digit stream by inserting predefined numbers, and select the trunk based on a wide variety of variables. The tables that define the routing sequence of calls in PABX are similar to those used by SS7 to route PSTN calls through the network.

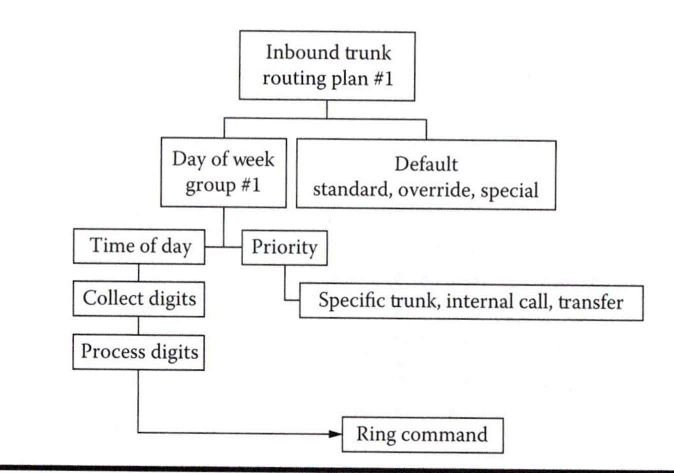

Exhibit 5.13 ACD programming flow.

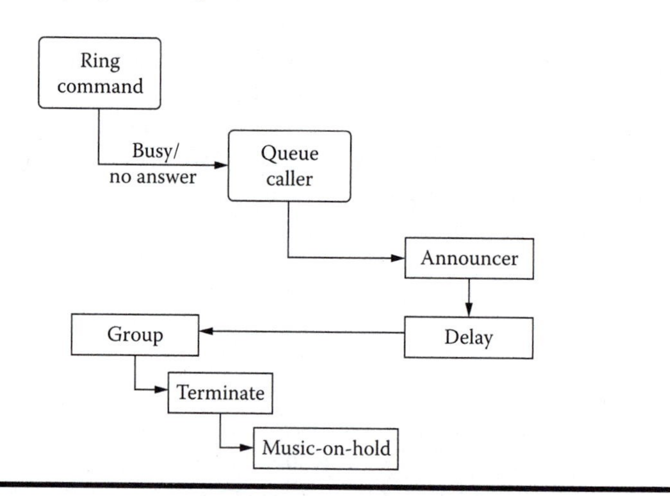

Exhibit 5.14 ACD call flow after ringing into an assigned group.

The advent of digital PABX gave business users sophisticated features years before the COs were upgraded to fully digital systems. Features such as call forwarding and call waiting were in place on PABX in the late 1970s, while most of the same class of service features did not appear in residential service until the mid- to late 1980s.

Digital Key Systems and Hybrids

Digital key systems come in various configurations, but they are consistently defined by the number of trunks and stations the control unit can support. A system term of 8/24 has the capacity for eight PSTN trunk connections and can provide access to 24 subsets. The control units of these systems are sealed units, which can be repaired by either the manufacturer or an authorized repair agent. The reliability of these units is extremely high, with mean time between failure (MTBF) rates comparable to CO equipment. However, when the system does go down, components cannot be easily replaced in the field, resulting in the entire control unit being replaced. The trade-off is a reduced cost for functionality over component-level replacement of hardware.

Hybrid systems can be configured as key systems with each inbound trunk available under a button on the subsets, or they can replicate the functionality of PABX with single-line subsets for the majority of users and a console at a main reception location. These systems have component-level configurations and are port-defined in their capacity. For example, a system that has 144 ports can have a maximum configuration of 128 stations and 16 trunks. The same system can be configured with numerous variations and combinations of trunks and stations, depending on the application demands. These systems evolved to replicate all the PABX features, including ACD, LCR, digital trunk interfacing, and other advanced system and station features.

This market segment is migrating toward a "soft" environment where the end user provides all the intelligence the system requires on an open system platform. These systems are configured on a nonproprietary operating system and have major components (e.g., power supplies) that are available from numerous suppliers. They have been based on PC platforms and have received negative publicity because of the lack of reliability or uptime compared to traditional telephone equipment. However, these new systems will eventually replace existing systems because they will be driven by end-user programming and not restricted to proprietary controls on processor upgrades or closed system architectures to peripheral devices.

Voice Processing

Voice processing can fall under an umbrella definition for two basic systems: auto attendant and voice mail (VM). The initial market representation of voice processing was characterized by those who held a strong disdain for interaction with a machine and those who enjoyed the increased productivity that the technology provided. Over the last three decades, voice processing has migrated to a highly technical process that can route and handle calls as efficiently as any human has in years past. Auto attendant has allowed the traditional receptionist to take on tasks other than being permanently connected to the main inbound call location. The ability of inbound callers to be routed directly to a designated department or individual with the auto attendant's assistance has changed the nature of call processing even for small businesses.

VM has as many supporters as it does detractors. The ability to save and organize messages without trying to decipher cryptic handwritten messages from a harried receptionist was a great leap forward in customer service. On the negative side, the same technology can be used to hide behind, as in cases where the called party never answers the phone, always allowing VM to screen all inbound traffic. The users (and abusers) of VM have transformed it into a commodity that we now enjoy with residential as well as cellular services. VM rapidly migrated down from being available only on larger PABX configurations to being fully compatible with small 3/8 digital key systems. A fully integrated connection, one that shares processor information between the voice-processing device and the telephone system, should be the only type of device installed for business applications.

Voice over Internet Protocol

Anyone who follows developments in the telecommunications industry is aware that VoIP is currently the topic du jour in telephony. For those who are just catching up, VoIP is the transmission of person-to-person voice conversations (which have traditionally been circuit switched) over networks that employ IP addressing. Over the past few years, the long-distance carriers (AT&T, Sprint, and Verizon) have been converting their national backbone networks to an IP-routed architecture. At strategic locations, Multiprotocol Label Switching (MPLS) routers are placed to connect routes across the network. Carriers are now basing their long-term corporate strategies on VoIP service to take advantage of MPLS routing technology. To remain competitive, the RBOCs (Verizon, AT&T, and Qwest) are offering VoIP service throughout their service areas.

VoIP Benefits

There are significant non-personnel-related savings to be realized via the merger of support functions. At a customer site, once the telephone, line, and PBX port are installed and associated, technicians must assist with the reconnection and reassignment of these facilities for moves and changes. With the advent of IP addressing for telephone equipment, much of this process has become automated as the IP telephone and enterprise server automatically exchange information among themselves once the phone is connected to the LAN. Most of the remaining telephone work is contracted out to a local telephone maintenance service. Furthermore, many existing lines can be eliminated, as the new IP phones connect over the enterprise data network and as the LAN is connected to each employee's workstation and IP phone. Then, courtesy of Dynamic Host Configuration Protocol (DHCP) servers, IP addresses are dynamically downloaded to the IP phone. These combined voice and data enterprise networks are scalable, users can be relocated, and offices can be restructured with much less administrative and rewiring expense.

Once a carrier's network is in place and support people are operational, it matters little to the carrier whether their distance network transports voice calls (originated as either analog or digital) or data traffic. All voice calls are converted into digital form at the local CO, and they are then transmitted through the network in digital form. All media, be it voice, data, or video, is carried through the network in digital IP packets.

VoIP Challenges

VoIP utilizes packet switching, which means it uses data networks to transmit the voice data packets. With packet switching, the connection is open only for the duration of the packet's transmission. This is an important concept because it is a way to transmit data without tying up the network with a single transmission's data. There are, however, a few issues that accompany VoIP and packet-switched environments. VoIP is latency driven, which means that it cannot tolerate any delay in the packet delivery in the transmission, so the network infrastructure must be able to provide uninterrupted delivery of the voice data, similar to that which is accomplished in a circuit-switched environment, with a similar high level of QoS.

Latency

Unlike most data transmitted IP networks, VoIP is latency sensitive, which means that latency cannot be tolerated in VoIP communications. Latency is the delay that occurs when a packet crosses a network connection. While other applications may not suffer from small delays, it becomes a major issue with VoIP. Many consider 250 ms to be the maximum allowable time for delay of a packet in a VoIP network. To achieve high-quality voice, the maximum desired one-way latency is 150 ms. If round-trip delays exceed 250 ms, users will notice the delays by the "clipping" of words and begin talking over each other. A call with latency above 500 ms is considered impractical.

Another component of latency is jitter. Jitter is defined as a variation in the delay of received packets and is usually caused by congestion in the network. At the sending side, packets are sent in a continuous stream with the packets spaced evenly apart. Due to network congestion, improper queuing, or configuration errors, this stream can become uneven, or the delay between each packet can vary instead of remaining constant. Packet delay occurs in the routers and is dependent on configuration, capacity, performance, and load. Routers also introduce around 10 ms of latency as they inspect each packet for delivery. To minimize this, IP telephony ports have taken priority over data ports. Packet loss also occurs in the routers. If the router is overloaded, it will choose to abandon certain packets based on the packet's priority label. The quality of a conversation will deteriorate if packet loss reaches more than 5%. If there is more than 5% loss, the performance of the call will be unacceptable.

Protocols Used in VoIP Environments

There are a number of protocols used in the transmission of VoIP services. The most common is the Session Initiation Protocol (SIP). Real-Time Transport Protocol (RTP) and MPLS are other common protocols that can be used in conjunction with SIP.

Session Initiation Protocol

SIP is what controls the setup, signaling, and teardown of communications through the packet-switched networks like the Internet. This can include voice and/or video calling, or instant messaging. SIP works on a client/server model, meaning there are many clients (i.e., smartphones, computers, tablets) that talk to one server. The server then talks to other servers, which help facilitate a connection between two clients. The components of this system are explained in more detail below.

Components of SIP

Uniform Resource Indicators

SIP clients and servers are identified by uniform resource indicators (URIs). URIs look like e-mail addresses. An example of one would be *sip:bob@business.com.* The first portion, *bob*, is the client name and is unique on the network or domain. The second portion, *business.com*, denotes the domain name. This would change depending on the network SIP is running on. For instance, a SIP client on AT&T's network might be *att.com* or *attindiana.com*. The Domain Name System (DNS) can be used to translate a standard telephone number to a URI. If a user were to type *1-765-867-5309* on their SIP-enabled phone, a DNS server would then translate that number to a SIP URI, like *bob@business.com*. The URI can be used to dial other SIP clients, but the historical use of telephone numbers makes them easier to use over URIs.

User Agents

There are two types of user agents (UAs): basic and advanced. A basic UA can perform functions of a feature phone, including call holding and transferring. Usually, basic UAs only support voice communications. Advanced UAs can support multiple media formats, including video. They are also capable of advanced features, such as conference calling and dialing using URIs or standard telephone numbers.

SIP Servers

Many of the initial call setup and tear processes are handled on behalf of UAs by different types of SIP servers. There are servers that keep a collection of UAs' names and addresses (registrar servers), servers that help communications between UAs in the same domain (proxy servers), and servers that help communications between different domains (redirect servers).

Proxy Servers

SIP proxy servers are used by UAs to communicate to other SIP servers and services. They act as representatives of the UA. The proxy server will connect to the registrar server for information about other UAs, redirect servers to find a UA, and find other proxy servers to initiate a call. The proxy server handles all communications on behalf of the UA until the UA establishes a connection to another UA. At this point, the proxy server can step away and let the voice call commence.

Registrar Servers

UAs reach out to registrar servers when they first join the network. They tell the registrar who they are, what their URI is, and other relevant information. The registrar server then takes that information and creates an entry for it in a database, or updates existing data in the database. A proxy server will ask a registrar server for information on where to find a UA when trying to establish a call.

Redirect Servers

Redirect servers help proxy servers and UAs locate other UAs on different domains. When a call is trying to be placed to a UA on a different domain, the proxy server will reach out to the redirect server on that domain and ask where to find the specific UA.

SIP Call Flow

Now that the components of a SIP system have been identified, we can put them together in one logical flow from the initial call placed by a UA up until the call is disconnected.

Exhibit 5.15 demonstrates a simplified SIP call process. In step 1, user A has picked up his/her receiver and dialed user B's number. The INVITE message is sent and rings user B's phone. In step 2, the RINGING message rings user A's phone, an indication that contact with the other UA has been made and that user A is now waiting for user B to answer his/her phone. When user B picks up the phone, in step 3, the OK message is sent to user A. Once user A's client sends an acknowledgment (ACK) message (step 4), the conversation can begin (step 5). In step 6, user B terminates the call by hanging up, which sends the BYE message to user A. User A then hangs up the phone as well, and in step 7, another OK message

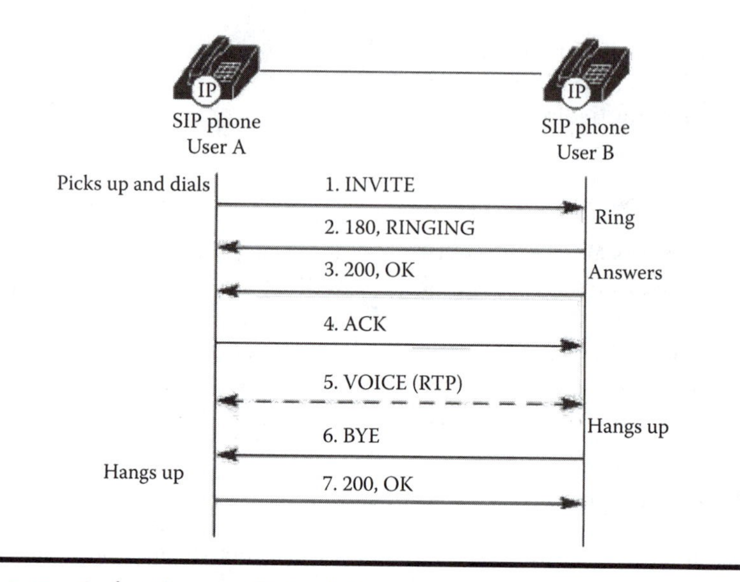

Exhibit 5.15 Basic SIP-controlled call flow.

is sent, confirming that the conversation is over and voice transmissions between the two phones can end.

Exhibit 5.16 shows the more complex process that a SIP proxy server goes through before connecting two UAs. Step 1 shows the UA registering with the registrar server. This step is usually completed when the UA is connected to the network. Step 2 shows the INVITE step from Exhibit 5.15 but in relation to the SIP proxy server. The proxy server, on behalf of Frank's IP phone, tries to find the UA

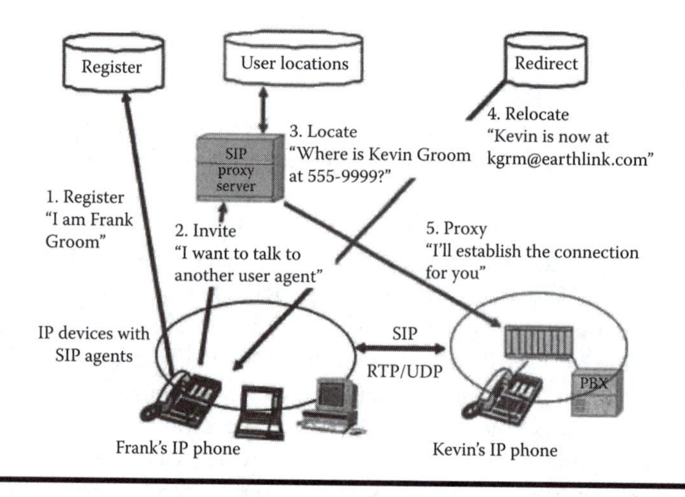

Exhibit 5.16 SIP servers and offered services.

with the number 555-9999. It looks for the UA in a database, and then a redirect server supplies the proxy server with a correct address. Finally, the proxy server contacts Kevin's IP phone and establishes the connection between Frank's and Kevin's phones. At this point, the INVITE step is complete, and step 2 from exhibit 4 begins.

SIP is currently the industry standard for signaling voice communications on the Internet and in private networks. Its distributed model of servers that work together to find and ring UAs is one of the most efficient ways to emulate the traditional PSTN so far. SIP's basic call flow model is also similar to how traditional phone systems work, the operation with which most users are familiar. With the use of URIs and mapping them to phone numbers, end users will not be able to tell the difference between a traditional phone system and a VoIP system using SIP. For now, VoIP in business and residential services will be handled with SIP.

Real-Time Transfer Protocol

When TCP is employed in addition to IP addressing, it can sometimes cause slight delays in the transmission. This is unacceptable because voice is extremely delay sensitive. When transmitting delay-sensitive traffic without using TCP, RTP is employed to supplement the connectionless best-effort User Datagram Protocol (UDP). Since there is no time allotted to waiting for the acknowledgment by the receiver, there is no time for the retransmission of the unacknowledged and presumably lost voice portions.

An RTP header is added to assist the receiving IP phone with potential difficulties, such as packets arriving out of sequence, arriving at irregular intervals, or needing to be buffered. RTP requires the overhead of an additional 12-byte header to accompany the UDP 8-byte header. The benefit of employing RTP is that it provides a time stamp for each voice segment that adjusts the pacing of received voice packets for the listener. RTP also provides a sequence number for reordering packets that might be delivered out of sequence. RTP helps the receiving device adjust the delivery by fine-tuning the timing of signal delivery to make it sound normal.

Multiprotocol Label Switching

Residential telephone service traditionally used analog circuit-switching protocols, but the core of the PSTN has been adapted to packet switching. ATM was adapted to work with IP technologies but soon proved to be inferior to a protocol called MPLS. MPLS works by analyzing a packet's header information when it is received by a label edge router (LER). The LER checks to see where the packet is heading and then determines where the packet's final destination will be. Finally, the LER places a label on the front of the packet and sends it to the appropriate label switch router (LSR). LSRs are devices inside an MPLS network that switch packets based solely on the label applied to them by an

LER. The packet is sent through the MPLS network until it arrives at its final decision, which is another LER. This LER strips the packet of its label and begins routing the packet according to IP standards, with source and destination addresses.

This entire process of applying labels to packets and switching them through the network has two main benefits. First, it allows many types of traffic to flow through a network without consideration of what the traffic is. Website traffic, e-mails, voice calls, and streaming video can all pass through an MPLS network without special configuration. MPLS supports traffic engineering, which can give priority to certain types of traffic, like voice calls. MPLS's second accomplishment is enabling an IP network to mimic some of the characteristics of the circuit-switched PSTN. It does this through traffic engineering, also called QoS. Traffic engineering in MPLS gives priority to certain types of traffic on the network by sending priority traffic first or discarding other traffic in favor of priority traffic. For example, it is vital to a phone conversation that all the packets arrive with minimal delay. Otherwise, the parties on the phone will notice that the quality of the call is poor. Compare that to loading a web page, where an extra second or so to access it can go virtually unnoticed. The traffic engineering function of MPLS is what makes it an ideal protocol for use in the core of voice networks.

Summary

When the PSTN was implemented, the goal was to provide universal access to telephone service across the nation. It seemed like a challenge at the time; however, this task was eventually accomplished and has had monumental impacts on society. The deployment of new technologies and the rise of more efficient methods of voice conversation are overtaking the services once dominated by the PSTN. Today's telephone services are increasingly routed through converged IP environments, leaving the PSTN behind. The growth of the Internet and the protocols that it uses to transmit information have given rise to telephone service provided exclusively through it. VoIP is becoming an industry standard for voice and data service.

Review Questions

True/False

1. A tandem switch provides dial tone to the end user.
    a. True
    b. False

2. RTP is used alongside TCP to ensure better voice quality.
   a. True
   b. False
3. Latency has little effect on VoIP calls.
   a. True
   b. False
4. VoIP is circuit switched.
   a. True
   b. False
5. PBXs serve as a routing device.
   a. True
   b. False

Multiple Choice

1. Which of the following is not associated with SS7?
   a. Service routing points (SRPs)
   b. Service control points (SCPs)
   c. Service switching points (SSPs)
   d. Service transfer points (STPs)
2. PRI stands for:
   a. Primary return interface
   b. Primary rate insert
   c. Primary ratio of an ISDN line
   d. Primary Rate Interface
3. Which of the following is a component used in SIP?
   a. SIP servers
   b. Proxy servers
   c. Registrar servers
   d. All of the above
4. A connection-oriented transmission is more reliable because:
   a. It is a best-effort transmission
   b. It uses a constant acknowledgment process between end points
   c. It cuts down on overhead, or management data, in the packet delivery process
   d. All of the above
5. Which of the following is a characteristic of packet switching?
   a. Dedicated connection between end points
   b. Originally designed for analog transmissions
   c. Breaks communication into chunks of data
   d. None of the above

Short Essay Questions

1. What does PSTN stand for?
2. What does VoIP stand for?
3. Give a name and define the out-of-band signaling technology discussed in this chapter.
4. How does ATM emulate circuit and packet switching?
5. What are the five main components of the PSTN?

Chapter 6

Basics of Multiprotocol Label Switching Networking

Within the telecommunications industry, there is an inexorable movement toward the universal employment of the Internet Protocol (IP) coupled with all other forms of networking protocols. From its Defense Advanced Research Projects Agency (DARPA) beginnings in the late 1960s to the 1990s Internet explosion, IP-based network transmission has been increasingly employed in corporate circles, publicly, and privately. So, our current fascination with IP is hardly a new phenomenon.

IP offers the important feature of multiplexing traffic from many users and integrating that traffic to be carried over a wide array of level 2 transport networks. Among these are Ethernet in the local area arena, Point-to-Point Protocol (PPP) over private lines, and frame relay and Asynchronous Transfer Mode (ATM) networks over the wide area. However, as users began working with a broader array of media types (including pictures, diagrams, video, videoconferencing, audio conferencing, and voice communications) over their IP-based networks, they became increasingly frustrated with the "best-effort" nature of IP's transport mechanisms. This is despite the fact that the IP suite of protocols and applications provides its own set of techniques to accommodate some of its difficulties with lost or delayed packets. Using Transmission Control Protocol (TCP) with IP, the user has the means to automatically resend lost or damaged packets. Also, using Real-Time Transport Protocol's (RTP's) timing and sequence numbering of packets, the packets that do arrive can be rearranged in the order in which they were originally sent and synchronized in timing as they are delivered to the receiving programs.

However, the *original* intent of IP's creators was to keep the protocol simple in operation, while distributing the intelligence required for its effective and robust operation throughout the network, hence the emerging need for better transport guarantees to meet customers' increasing needs for security, ever-higher bandwidth, bandwidth on demand, and control of delivery.

Introduction

It has become increasingly clear that for large public and private networks, customers were increasingly demanding throughput assurances concerning the level of service, as well as the speed and quality of delivery, that the traditional telephone network and most private networks could not provide. To satisfy this customer demand, many suppliers, such as Ipsilon with its IP-switching approach, Cisco with its tag-switching approach, Toshiba with its Cell Switch Router, and IBM with its Aggregate Route–Based IP Switching (ARIS), have attempted to find a remedy. Fortunately, all parties quickly recognized the importance of being able to pretag packets with an identifier, and they ultimately allowed their separate approaches to be merged into one standard approach, called Multiprotocol Label Switching (MPLS). MPLS retains many of the benefits of the previous approaches, while allowing all vendors to support a standard multivendor, multiprotocol approach.

What MPLS needed to position itself at the heart of almost every vendor's network strategy was an application that was universally appealing. In the late 1990s, such an application emerged as a result of the popularity of the public Internet. To complement this public Internet, many carriers and service providers offered "private" Internets to deliver service connections around and occasionally through the public Internet. The offered product was termed a *virtual private network* (VPN); it was a managed service, employing encrypted IP traffic to and from the user's sites and appearing to the user as if all traffic were transmitted over the public Internet. This was done regardless of the true path and facilities over which it was actually transported.

MPLS was initially employed in enterprise networks where guaranteed paths were required for very time-sensitive traffic, such as voice and video conferencing applications. However, with its attractive services, such as VPNs and customer control of bandwidth through a product termed *traffic engineering* (TE), MPLS is now finding a home in the national carrier networks. This is a result of many factors, among which are the following:

1. The carriers must significantly reduce their cost of provisioning, maintaining, reallocating, monitoring, and managing their national networks and the individual customer transmissions that flow over it.
2. Customers are demanding the same quality of service for their intersite transmissions that they previously experienced with their on-site local networks.

3. High-speed (broadband) access to the Internet is suited to IP routing rather than the carrier's traditional circuit-switched technology.
4. Competition has encouraged the emergence of lower-cost service providers who make use of incrementally growable (as their customer base increases) switching and routing facilities through which they can offer least-cost versions of common services. Moreover, these service providers must use technicians who have undergone common, universally available training. This has led providers to the employment of common multiservice and IP-based networks.
5. The deployment of fiber transport media in the local access, metro, and national backbone networks allows users to stream a variety of media types—increasingly, voice (Voice over IP [VoIP]) and video (IP video) applications.

As a result, there is now a significant movement toward MPLS implementation in the national carrier networks, in the local exchange networks, and in larger enterprise private networks. Moreover, the recently formed carrier/local consolidated service providers (AT&T and Verizon) have added MPLS to their national IP backbone networks as well as to their regional and metropolitan networks. With this incentive, we can now discuss the basics of MPLS and the varieties of its deployment.

What Is MPLS?

MPLS is a switching approach that is meant to implement ATM-like virtual-circuit-equivalent approaches into an IP overlay on existing networks. This virtual-circuit approach employs a set of routers that are applied as an overlay on the edges of the carrier network and a labeling scheme in which the label is added onto the addressing header of each carrier layer 2 frame being transported. The inserted label indicates to each router along a selected path through the network what quality and class of service should be provided to the media (frames, packets, or cells) that are transmitted across the network. These labels are used by the network to choose, among a set of virtual circuits, ones that best meet the customer's requirement. Such virtual circuits can be either precreated and waiting to be employed (termed *permanent virtual circuit* [PVC]) or dynamically created (termed *switched virtual circuit* [SVC]) in response to a "request" message submitted by the customer to the edge network node of the service provider's MPLS network.

This MPLS request is similar to ATM's call setup message submitted by an ATM network user, which requests that a certain quality of transmission service, meeting the user's requested requirement, be set up across an ATM network. Whereas ATM is a layer 2 switching technology, IP is a layer 3 routing technology. Both employ a table at each network node (either a switch or a router) to decide how to switch or route each cell or packet along a link to the next node in the selected path.

Under MPLS, layer 2 frames containing packets are forwarded on the basis of a label that is inserted in the frame between the frame address and the packet address as a response to the user's requesting message specifying a level of transport. The label indicates to the MPLS network which circuit these frames and enclosed packets are to employ. As mentioned earlier, this process is meant to operate in a fashion similar to ATM switching. MPLS's label-based forwarding process is performed by means of a table lookup based on the label that has been inserted in the packet. Once a match has been found for the label in the router's table, the packet is forwarded to the appropriate output port and relabeled for transfer along to the next hop in the selected path to the ultimate destination. As a result of how the MPLS labels are inserted in the packets and the process of label-based forwarding (a variation on switching and routing) by MPLS routers in the network, MPLS is considered a layer 2.5 protocol, lying halfway between layer 2 and layer 3.

The paths that MPLS networks use across the network are called label-switched paths (LSPs) and employ virtual circuits across the network.

Incoming customer traffic to the MPLS network is assigned to a specific class called a Forwarding Equivalence Class (FEC). This class is assigned to a specific path through the network, and all traffic to a specific destination and requiring a similar treatment is assigned to the same class (FEC).

Motivation behind MPLS

Customers as well as service providers employing the transmission of IP traffic over telephone circuits, private lines, and the wide area network (WAN) services of frame relay and ATM have experienced limitations. Customers have increasingly wished to dynamically request varying quality-of-service characteristics for individual traffic and have been frustrated by what has been available. Carriers have been equally frustrated in their ability to meet customer demand for such variability.

Employing MPLS as an overlay network surrounding the facilities of an existing carrier network allows for the use of MPLS as a forwarding system in which customer-submitted packets are forwarded along a designated path to the desired destination edge of the MPLS network based on the MPLS label that is inserted in the customer packet's header. As a result, adding an overlay network to the facilities of an existing network has allowed carriers a great deal of flexibility and control in the process of meeting customer needs over their existing facilities of the long-distance network while meeting a wide range of customer bandwidth and quality demands.

Many customer issues with distance services have been long-standing in nature. Among these is the wish of customers to have some control over picking paths for their traffic that can avoid congestion. Moreover, privacy of transmission is becoming an increasingly important requirement for both corporate and customer information as well as the individual transaction being transported. With the employment of MPLS surrounding the carrier networks, VPNs can be provided over these traditional carrier services and facilities. Furthermore, TE can be implemented to carve out required pathways through the network to meet individual customer momentary demands. Further, fast rerouting of traffic and preemption of lower-priority traffic allows carriers to deliver service with much greater flexibility. All this can be accomplished as well as the integration of existing WAN service offerings into the MPLS umbrella offering.

Carriers were finding it difficult to configure their underlying layer 2 ATM networks on the fly to meet customer on-demand requests. Furthermore, these underlying ATM networks forced a single structure for the serving network, and the kinds of service ATM was designed to provide and the way it provided these services limited what could be offered and how it could be offered.

Furthermore, the difficulty in recruiting and training ATM engineers placed an additional burden on the carriers, while the cost of ATM equipment soared as the demand for high bandwidth resulted in faster and faster turnover of the switching, interface, and optical connecting equipment. ATM had a high level of overhead (commonly termed the *ATM tax*), which many vendors were interested in minimizing. To guarantee a consistent quality of service, ATM requires that the network conform to a strict set of standards, a limited set of services, and only a few sets of interfaces.

On the other hand, despite its flexibility, IP still tended to deliver a slow table lookup process based on a destination address at each hop, resulting in a sequence of hop, table lookup, hop, and table lookup at each step to a destination. Whereas layer 2 switching speeds were improving rapidly, IP routing was only gradually improving. MPLS forwarding based on inserted labels emerged as an attractive compromise solution.

Some Attractive Features of MPLS

IP-based networks, such as the public Internet, are prone to congestion and slowdowns. Furthermore, routing of IP packets is a slow and time-consuming process. The MPLS approach forwards submitted packets on the basis of a label that is inserted by the entry at the edge of an MPLS network. This process is efficient and quick, while removing much of the processing and route optimization from the MPLS routers (termed *label switching routers* [LSRs]) at the heart of the network, and places the responsibility for path optimization at the edge with the label edge router (LER), which initially receives the customer's packets.

MPLS packet forwarding is accomplished in a similar fashion to ATM cell transmission, whereas MPLS path reservation and label distribution are also performed in a manner similar to ATM. Furthermore, both MPLS and ATM disseminate path information in a variation of IP's open-shortest-path route dissemination procedure. However, although routing protocols are used to set up MPLS paths, the routers along the selected path do not route on the enclosed destination IP address but, rather, ignore that address, and an MPLS mechanism in each router forwards the packets based on the added MPLS label. MPLS, furthermore, similar to ATM, signals to the routers along the chosen path that certain *in* ports and *out* ports will be used as part of a designated MPLS path.

As a result of surrounding the carrier's ATM core network with an overlay MPLS set of routers (or even over a large enterprise Ethernet network), the performance of the network in carrying a mixture of customer traffic at a variety of quality levels is significantly improved, while the overhead and slow processes of IP routing are avoided. Furthermore, carriers such as AT&T and Verizon can now maintain a broader level of control of both the overall network and individual customer requests and traffic flow.

Principal Components of an MPLS Network

A number of terms describing devices employed in an MPLS network need to be defined, and a few of the mechanisms used have to be described before we discuss MPLS functions. We will need a few of these definitions to proceed. Among these components are the following:

- Customer edge router (CER): The customer's router at the edge of the customer's network that submits both requests for MPLS paths (and labels) and frames and packets for subsequent labeling and forwarding over the MPLS network.
- MPLS label: A header placed in the address field between the layer 2 and layer 3 addresses that is used by the MPLS routers in the network to switch/route the frames and packets.
- Label edge router (LER): The router at the entry point to the MPLS network. This router receives customer request messages, applies an appropriate label to indicate the path that provided that transmission quality, and advertises that that label and path have been subscribed to.
- LSR: A router within the MPLS network that switches/routes that message on the basis of the MPLS label that has been inserted between the level 2 and level 3 addresses.
- FEC: An FEC identifies a particular traffic flow along a path through the network. That preassigned path is designated an LSP through the network, and each FEC class is assigned to a particular LSP path. All customers whose packets are assigned to a particular FEC class have their traffic flow along the same prearranged LSP path.
- LSP: A chain of labels that, in sequence, are applied to the message as it is forwarded from one router to the next along the designated path associated with the required bandwidth and service quality.

- Label Distribution Protocol (LDP): The protocol employed by the entry LER to notify the downstream routers that a certain pathway has been allocated and that they should place labels in messages passing through along that pathway.
- Label Information Base (LIB): The base of information about a set of labels defining a chosen path that is forwarded to the LSRs along the chosen path.

Operation with Edge and Core Label-Switched Routers

As previously mentioned, MPLS operates in a halfway house between layer 2 switched networks, such as Ethernet, frame relay, and ATM, and the layer 3 world of IP routing. Layer 2 networks operate link to link from one network node to another. Each layer 2 address addresses only the next node in the chain, not the final destination. Layer 2 networks employ switches at each node. IP, the most popular layer 3 network, employs routers at the nodes (instead of switches) and makes forwarding decisions on the basis of a constant destination IP address contained in the header area of each packet that arrives at the routers. Exhibit 6.1 portrays such an MPLS network interconnecting an "originating" and a "terminating" network. The underlying layer 2 mechanism of the MPLS network is not shown but merely implied.

The routers at the entry and exit edges of the MPLS network apply the labels to the customer-submitted packets that arrive from other networks to the MPLS core network. These edge routers must first communicate with all the other routers in the MPLS network to receive and understand the topology of the overall network, including information about all of the links as well as the capabilities of the network and any prior commitments to individual links.

The entry routers also pick a set of labels, which can subsequently be used to address packets so that they can take particular routes across the network that conform to a particular class of throughput and quality of performance.

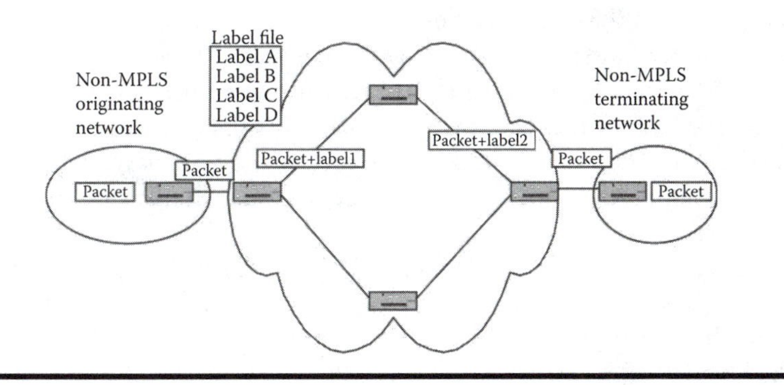

Exhibit 6.1 Labels applied to packets as they arrive at the MPLS network.

To begin transmitting a customer's submitted packets, the LER at the entry point to an MPLS network first receives customer request messages. The path these packets are to take through the MPLS network can be predefined as a PVC through the network, or a path can dynamically be set up in response to a customer's request for a specific bandwidth and service level at that moment. Regardless of which approach is chosen, the MPLS edge router (LER), which receives the customer's packets, then places a label in the "shim" location between the layer 2 and layer 3 addresses in each submitted customer packet. The group of packets that are headed for the same location follow the same path and are to receive the same treatment across the network; these are placed into a special group termed an *FEC*, where all will be forwarded in a common manner, over the same path and with the same treatment. At each router along the way, the label of the received packer is removed, and a new label is added. This process is termed *label swapping* because a new label is employed to traverse each link across the path of routers through the MPLS network. At the exit point, the last label is removed, and the shim location is removed from the header area of the customer's message.

Labels and Label-Switched Routers

MPLS employs a 32-bit label that is added to each transmitted frame. This MPLS label can be inserted between the layer 2 and layer 3 addresses (commonly termed *frame-mode MPLS*) or it can be substituted for the layer 2 address itself, as employed when ATM cells are transmitted over an MPLS network (commonly termed *cell-mode MPLS*).

The special MPLS labels are applied to the packets (inside layer 2 frames) by the edge device when a frame carrying a packet enters the MPLS network from a non-MPLS network. This non-MPLS network is typically an enterprise site network employing Ethernet or a WAN service employing frame relay or ATM. At the other end of the MPLS network, the special MPLS label is stripped from the packet/frame, and the frame carrying the packet is forwarded on to the destination delivery network.

Within the MPLS network, both the layer 2 and layer 3 IP addresses are ignored. Only the inserted MPLS label is examined. At each step, the entering packet's MPLS label is compared to a table, and the proper output port is determined on the basis of the table entry. The packet is then switched to that output port, and a new MPLS label is added to the packet (and a new layer 2 frame address is added as well). This process is termed *label swapping*. Exhibit 6.2 portrays this swapping process.

As labeled packets are forwarded through the MPLS network, each router along the way swaps out the label in the incoming packet with a new one for the next hop along the chosen path.

The function of the entry point to the MPLS network, the LER, is to (1) receive the customer's request for service, (2) make a determination as to how the request

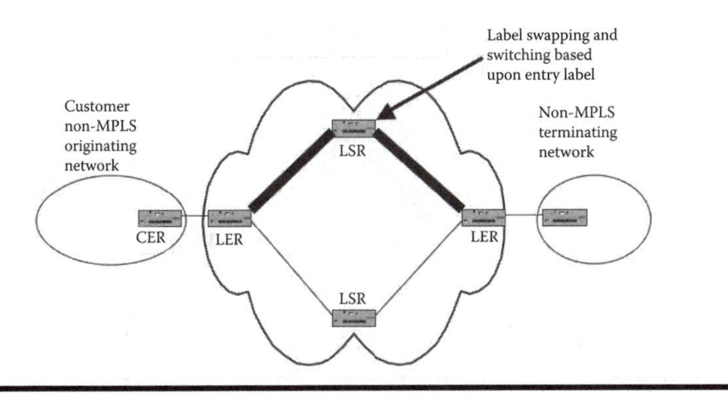

Exhibit 6.2 Label swapping as packets are forwarded through an MPLS network.

can be satisfied by the network and how the packet will traverse the MPLS network, (3) assign the appropriate label (either as a shim label or by replacing the ATM cell label), and (4) pass the packet on to the next LSR in the determined path.

The entry router employs a routing protocol to determine the reachability of the requested destination network. This LER determines not only reachability but also destination mapping and then labeling. It then creates a set of labels describing a particular path and then employs one of a set of possible distribution protocols (LDP or Reservation Protocol as used for traffic engineering [RSVP-TE]) to distribute these labels and mappings to the routers along the associated path. Subsequently, this entry LER adds a label to each packet and forwards them along the determined path.

Forwarding Packets with MPLS Labels

Each router along the MPLS network maintains a forwarding table. This table contains an entry for the input port and the label expected for packets that arrive at that port. It also contains a number of output port entries and the label to be applied to traffic leaving that port.

The next router (the LSR) along the determined path receives the labeled packet, strips off the layer 2 address, determines the next hop to be traversed on the basis of the label in the shim part of the packet's header, assigns a new MPLS label in the shim part of the header, switches the packet depending on the enclosed label to the appropriate output port, adds a new layer 2 address of the next LSR router in the path, and then passes the packet along to the next LSR router in the path.

The final LSR (another edge LER router) strips away both the layer 2 address and the shim MPLS label, and places another layer 2 address in the header for transmission over the destination layer 2 network. No further MPLS addresses are needed. Exhibit 6.3 presents this forwarding process, which is based on the inserted labels.

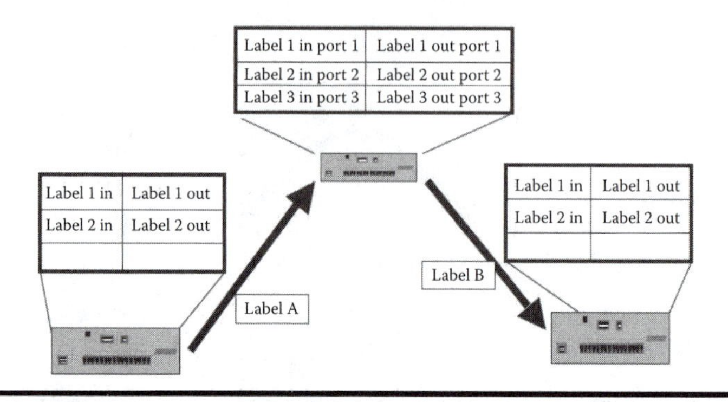

Exhibit 6.3 Using label-forwarding tables.

Example of MPLS Label Swapping

In the following example, a packet with the IP address 125.89.25.6 enters the MPLS network (Exhibit 6.4). The LER assigns a label of *3* and forwards the frame out through output port 1 to the circuit attached to that port.

The next router along the path, another LSR, brings the packet in port 2 and, depending on the contained MPLS label of *3*, forwards the packet (relabeled *3* again in this case) out through port 3 to the connected circuit. The last LER brings the packet in and, depending on the contained label *3*, forwards the packet, which is now exiting the MPLS network, to the connected layer 2 delivery network.

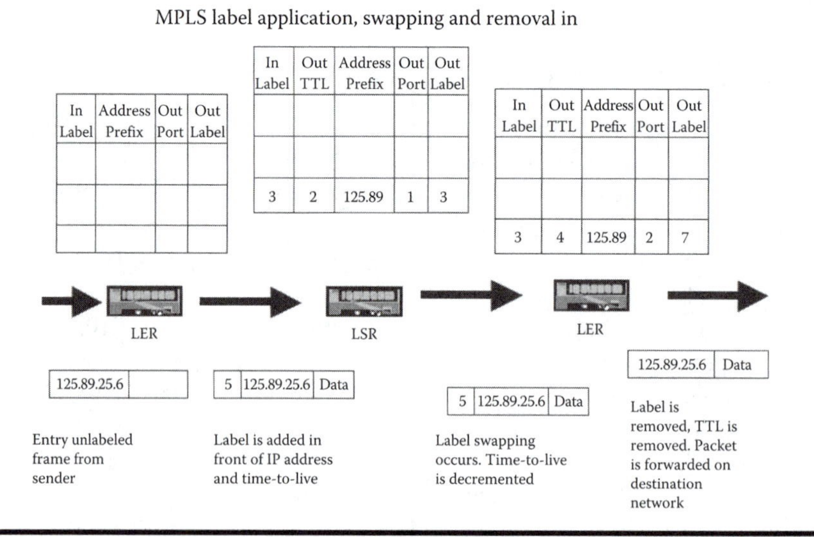

Exhibit 6.4 An example of an MPLS labeling, label swapping, and forwarding process.

Label Switching

Labels

MPLS is a forwarding process in which received packets are forwarded on the basis of enclosed labels. Labels are placed in the header of the packet by the MPLS entry LER router. These labels can correspond to destination addresses, levels of quality of service, or just the required transmission speed of the selected path.

MPLS is designed to forward a wide array of protocols—not merely IP packets. However, most layer 2 protocols such as Ethernet in the local, and increasingly in the metro area, and frame relay and ATM in the wide area, are populated by the carrier through an MPLS network with the path designated by the embedded MPLS label rather than the layer 2 address or encapsulated IP packet's address. Delivered packets using other layer 2 protocols such a PPP and HDLC can also be forwarded through an MPLS network.

Labels and Label Edge Routers

Labels are placed in the header of the packets by the MPLS entry (LER) router. These labels can correspond to destination addresses, levels of quality of service, or just the required transmission speed of the selected path. The logical path that the packet containing an MPLS label follows is considered a label switch path (LSP). A group of packets, all requiring the same treatment, headed for the same destination, and all following the same path through the MPLS network are considered to be forwarded in a shared equivalent manner and thus are members of a shared FEC.

Labels and Shims

MPLS is a transmission protocol that makes a path selection decision on the basis of a label inserted into a framed packet that is to be transported. The MPLS label identifies one of a number of possible forwarding classes to which a set of packets is to be assigned. Each forwarding class describes a level of transmission service required to reach a specific destination edge router in the MPLS network. The inserted MPLS label consists of the 20 bits identifying the FEC to which the packet is assigned, an unused experimental field, a bottom-of-the-stack indicator, and a time-to-live (TTL) field. The TTL field gets decremented each time the packet goes through either an LER or LSR router. When TTL reaches 0, the packet has been circulating too long in the network and is discarded as undeliverable. Exhibit 6.5 shows the MPLS label format.

It would be nice to say that these MPLS labels are always applied in the same way to all frames and packets that are to be forwarded over the MPLS network. However, there are actually two ways to apply these labels. The first way, the shim way, is to insert the MPLS forwarding label indicating the path the packet will follow, between

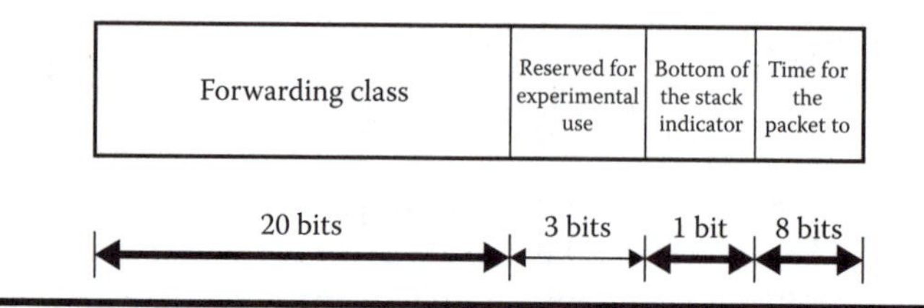

Exhibit 6.5 The MPLS label to be added to each packet for forwarding.

the layer 2 header (address) and the layer 3 header (address). The MPLS label is positioned as a shim between the other two addresses, which the packet already contains. These layer 2 and layer 3 addresses will be used at the destination network. However, only the MPLS forwarding label is used to traverse the MPLS portion of the network. Exhibit 6.6 shows this MPLS shim added to the packet's header.

The shim approach is employed when transporting IP packets (layer 3) that are contained within Ethernet frames (layer 2) when MPLS is used in a local Ethernet-based enterprise network. The shim approach is also used over private lines such as leased T-1 trunks, in which a layer 2 (PPP) header is employed or when High-Density Link Control (HDLC) is employed as the point-to-point addressing over a T-1 or DS-3 circuit.

Frame relay service is probably the most popular layer 2 service available to provide a managed transmission over a regional or national carrier network. When forwarding frame relay frames over an MPLS network, the MPLS label is placed as a shim between the frame relay layer 2 header and the IP layer 3 header, much similar to the process employed for Ethernet in the local area. Exhibit 6.7 shows such a label insertion.

However, MPLS uses a different approach for applying the MPLS labels to ATM cells. The ATM approach takes advantage of the fact that ATM applies virtual-path identifier (VPI) and virtual-circuit identifier (VCI) addresses at each hop along a designated ATM path to a destination ATM edge switch. This path can be based on a virtual circuit that has been preestablished (termed a *PVC*) and available to the entry ATM switch, or the desired path can dynamically be selected (termed an *SVC*)

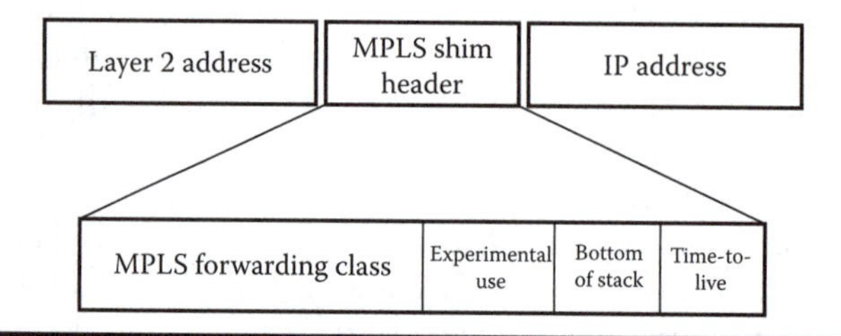

Exhibit 6.6 The placement of the MPLS shim-forwarding header.

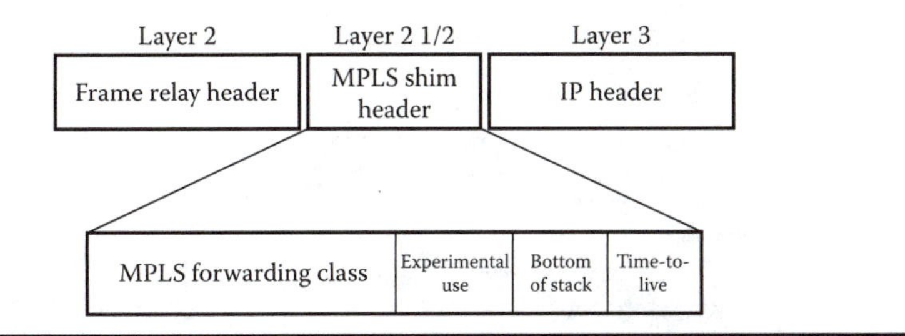

Exhibit 6.7 Forward label applied in shim to frame relay header.

depending on a customer request message. ATM's addressing scheme is similar to the MPLS approach. MPLS takes advantage of this similarity, and instead of creating a shim address between ATM addresses, it overlays the ATM virtual-path and virtual-channel addresses of each consecutive cell with a successive set of MPLS labels at each hop along the chosen path—much as ATM would insert new VPI and VCI addresses as cells proceed switch by switch through an ATM network.

Not just 1 but, rather, a sequence of 30 ATM cells are used to carry one IP packet. The first ATM cell carries the ATM address of the virtual path and virtual channel, whereas the IP address for the carried IP packet is carried in the data portion of the cell. The subsequent 29 ATM cells carry the same ATM address, but each cell carries a 1/29 fragment of the original IP packet data. Exhibit 6.8 presents the first ATM cell with the IP header address, whereas the 29 following cells contain the data fragments.

When these 30 ATM cells containing one IP packet are carried over an MPLS network, the ATM address is replaced with the assigned MPLS labels for forwarding over the MPLS network. No shim address field needs to be installed between the layer 2 and layer 3 address fields. The layer 2 ATM address is overlaid with the MPLS label because the virtual-path address and the MPLS label serve the same function as a link-by-link swappable address locally applicable to just the link and not the destination location.

Exhibit 6.8 ATM cells with MPLS label shim replacing VPI and VCI.

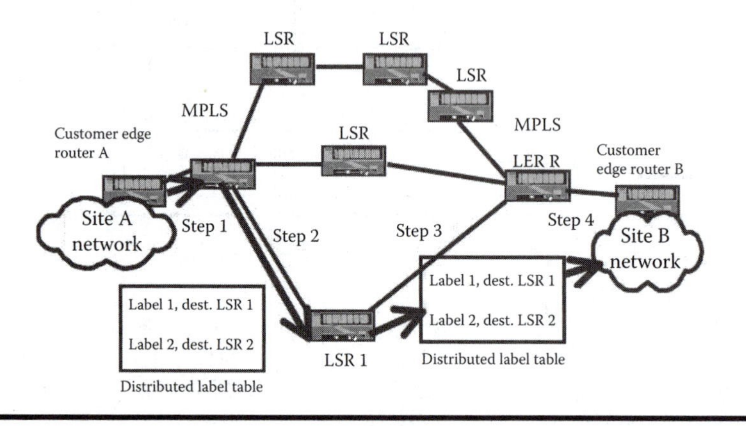

Exhibit 6.9 Distributing the labels that are to be used along a particular path.

This set of ATM cells and overlaid ATM addresses with MPLS labels is depicted in Exhibit 6.9.

Signaling and Label Distribution Protocols

Exchanging the routing and label swapping information, and the FEC classes and LSP paths that the FECs are assigned to, is considered part of the control plane of the MPLS network because they are the specifications and mechanisms by which traffic flow is "controlled." The packets to be forwarded, the edge LERs and core LSR routers, and the links between the LSRs are the physical part of the network and are considered the physical data plane of the network.

The signaling protocols that MPLS edge routers (LERs) use to establish a path through the network that satisfies a particular class of service (an FEC) are limited to two protocols, with a third as a variant on the second.

Network Topology Discovery

The most common approach followed by many types of network protocols (such as ATM, frame relay, and IP) is to have the nodes in a network periodically exchange their common view of the network. The Open Shortest Path First (OSPF) approach is a common information exchange protocol followed. Under OSPF, the routers periodically transmit their tree structure of the network. With that exchanged information, each router builds its own view of the links and capabilities of the network in a common fashion. In MPLS, a similar exchange occurs. The available paths and the capabilities of each link are forwarded between routers, and subsequently, a label associated with each link is distributed by the entry LER.

Thus, MPLS allows for the distribution of the information concerning the overall structure and topology of the MPLS by means of well-known routing protocols. These protocols can be employed to distribute information between the MPLS endpoint routers (LERs). Routing can also be used in the initial setup of the router-based network. MPLS even allows for the further overlaying of non-IP-based traffic flows (although IP traffic is the predominantly carried traffic) onto the LSP-based network.

Reservation Protocol

The most widely implemented signaling protocol is RSVP-TE. RSVP has been employed by many systems that guarantee transmission quality of service that meets customer requests. Generally, all vendors support RSVP for MPLS signaling, which makes RSVP a popular choice. With this protocol, a request message is sent along the pathway that the LER determines to be most appropriate for this FEC class. This request message provides the information that the path is available to the LER at the other end. Then, the furthermost router (the LER at the other end of the network) sends back along the pathway a specific message requiring that all the LSRs along the way reserve that path for traffic assigned to the designated FEC.

Label Distribution Protocol

This exchange mechanism is called the LDP and operates similarly to OSPF's information distribution mechanism. This commonly used OSPF routing protocol falls within the general classification of Interior Gateway Protocols (IGPs). Thus, when a single MPLS network is to be traversed, the LDP can be employed to distribute the labels in a fashion similar to IP's OSPF Interior Gateway approach. Border Gateway Protocol (BGP) is employed when distributing across the bridge between two networks.

LDP has been designed specifically for MPLS and incorporates many TCP-like facilities within the signaling protocol. For example, each router in the network sends a "hello" or "keep alive" message periodically to its neighbors to "assure" them that they are still active and a path remains available. LDP is a protocol that executes hop by hop in an IP-like fashion and utilizes the same path for traffic that the IP route distribution dissemination protocols such as OSPF use. At the moment, LDP does not support the customer's ability—termed *TE*—to engineer on demand a specific path to be employed as traffic is submitted. Fortunately, RSVP-TE is available to satisfy such a requirement and is generally available.

When TE is also available to customers of the network, RSVP is employed whereby the customer sends a "path reservation" message to the network first to institute the pre-engineering of the bandwidth to be used subsequently by the submitted packets. In this case, RSVP, rather than the LDP approach, is also used to distribute the labels within the router. The other frequently offered MPLS service (VPN service) is independent of the two methods to distribute labels and can employ either method.

The special signaling protocol MPLS employs to distribute labels throughout the MPLS network generally follows one of two forms. With the more common form, termed *downstream on demand*, the upstream entry router (LSR) requests of downstream routers that they assign a label to a particular class of traffic—an FEC. The downstream routers then notify the entry router of the labels they have assigned for that class, and the labels are bound to that class of service; there is a common understanding among the routers along that path that all packets of that FEC or class of service will follow.

With an alternative method, termed *downstream unsolicited*, when routers downstream in the network discover a new router that has capacity that might be used in a particular FEC class, they can apply a label to a particular next hop and notify the upstream routers that that label and that router and path are available to serve that class of traffic. The upstream routers then insert that label in their forwarding table to use when indicating the new path alternative. Exhibit 6.9 portrays traffic from site A to site B and the distribution of labels to enable the forwarding of submitted packets across the MPLS network.

Generally, the entry point to the MPLS network, the LER, picks the path that traffic of a particular class of FEC will follow and the labels to be employed, and the downstream routers, LSRs, comply. This label distribution process is illustrated in Exhibit 6.10.

Constraint-based routing label distribution protocol (CR-LDP), a variation on LDP, has been issued by the ITU as an international signaling standard for MPLS

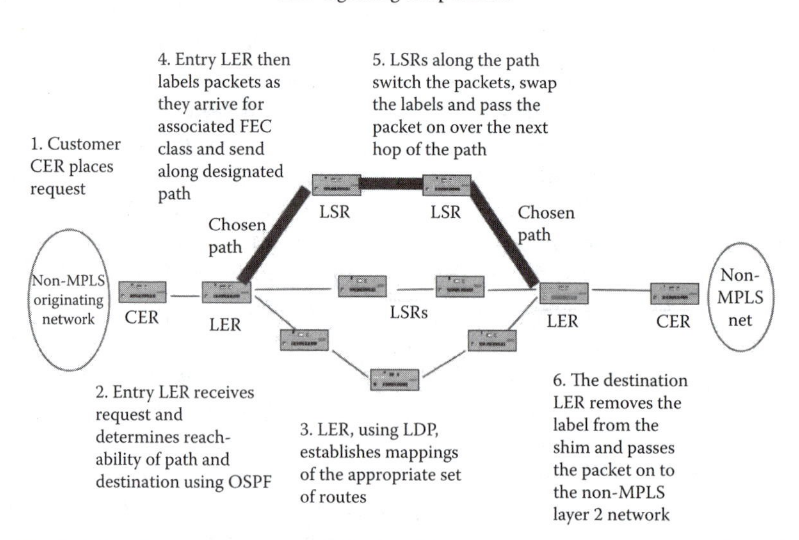

Exhibit 6.10 Label Distribution Protocol (LDP), packet forwarding, and label swapping in action.

that also supports TE (the satisfaction of customer bandwidth paths on demand), although it is relatively unused currently.

Exhibit 6.10 portrays the six-step process for satisfying a customer request for dynamic bandwidth across an MPLS network.

Forwarding Equivalence Classes and LDP Paths

Although many MPLS networks support on-demand satisfaction of customer bandwidth requests, the more general approach is to precreate a number of paths at varying levels of bandwidth through the network. Customers who request service that utilizes one of these paths across the MPLS network are assigned to a particular class or FEC that is associated with one of those prearranged paths. Therefore, each FEC identifies a particular traffic flow along a path through the network. That preassigned path is designated an LSP through the network, and each FEC class is assigned to a particular LSP path. All members of a certain FEC class have their traffic flow along the same prearranged LSP path. The labels to be assigned by each successive LSR along the path are distributed by the signaling protocol: either RSVP or LDP. Packets are assigned to a particular FEC when they enter the MPLS network through the entry LER. Forwarding of the packets then follows the LSP path set up for the particular FEC that was preassigned to a path through the MPLS network. Individual packet forwarding then is based on the label that is successively placed in the shim of the packet header at each hop along the preassigned path for that FEC class of traffic.

So, an FEC can be considered a group of packets that are to be forwarded along the same path through the network and receive the same treatment at each hop along that path as successive LSR routers determine the next hop to forward the packets to based on the FEC class that each packet is assigned to and the label signifying that FEC that has been inserted in the shim part of the packet header.

An LSP is an end-to-end virtual circuit similar to an ATM network virtual circuit. Many LSPs may traverse a particular link of a path through the network. Thus, a particular packet has inserted in its header a particular label that is used to identify a particular LSP path over which that packet, and in fact, all packets, assigned to the same FEC class must pass.

Each successive LSR router along a path uses the label in the header of the incoming packet, checks the Next Hop Label Forwarding Entry, and then relabels the packet for the next hop along the preassigned LSP path for its assigned FEC class.

Paths that an LSP can be assigned to can be identified in a hop-by-hop fashion whereby the LSP router looks up the label in an IP routing table to find the next hop router along the path and the appropriate label to insert in each packet, or the particular nodes in the path can be predesignated.

When configuring the LSP path, you can manually configure the incoming and outgoing labels and associated ports in a router's label swapping table (routing table) in a fashion similar to manually preestablishing a PVC for an ATM and frame relay networks. Alternatively, you can employ one of the signaling protocols previously discussed, especially RSVP-TE or LDP.

Conclusion

MPLS has proved quite useful as it is employed by national carriers to carve out paths from its network and telephone facilities that meet customer requests and expectations on a consistent basis. Employing a set of routers that are overlaid on the telephone long-distance network and a set of MPLS protocols that specify particular classes of traffic, MPLS provides a great deal of control over traffic flow to both the carrier and the requesting customer whose traffic is transported. A number of services are currently provided that meet customer demand. MPLS is employed to deliver encrypted VPN services that mimic the security and assurance of a private point-to-point circuit, and customers can also request and control the amount of bandwidth they employ for selected traffic by subscribing to a TE service over the offered MPLS network. As increasing portions of the nation's traffic are IP based, MPLS technology can be overlaid to enhance the ability of service providers to meet customer demand while they modernize their existing networks and move consistently toward fiber-based delivery of IP voice, data, and video traffic over a common set of carrier facilities.

Review Questions

True/False

1. The label edge router (LER) is the customer's router at the edge of the customer's network that submits both requests for MPLS paths (and labels) and submits frames and packets for subsequent labeling and forwarding over the MPLS network.
   a. True
   b. False
2. A label switching router (LSR) is a router within the MPLS network that switches/routes that message based upon the MPLS label that has been inserted between the level 2 and level 3 addresses.
   a. True
   b. False

3. An MPLS label is a header placed in the address field between the layer 2 and layer 3 addresses, which is used by the MPLS routers in the network to switch/route the frames and packets.
   a. True
   b. False
4. MPLS is considered as a layer 2.5 protocol.
   a. True
   b. False
5. MPLS places the labels in the shim except for ATM cells.
   a. True
   b. False

Multiple Choice

1. MPLS stands for:
   a. Multiprotocol Level Security
   b. Multiple Protocol Labels Service
   c. Multiple Protocol Labels Switching
   d. Multiprotocol Label Switching
2. Which is responsible for path optimization?
   a. LER
   b. LSR
   c. CER
   d. Switch
3. The LSRs read/change:
   a. The MPLS label and the layer 2 address
   b. The MPLS label and the layer 3 address
   c. The MPLS label and both layer 2 and layer 3 addresses
   d. The MPLS label only
4. The MPLS signaling protocol is based on:
   a. RSVP-TE
   b. LDP
   c. Network topology discovery
   d. All of the above
5. MPLS employs a:
   a. 2-byte label
   b. 32-byte label
   c. 4-bytes label
   d. Variable-size label

Matching Questions

Match the components of an MPLS network to their proper description:

a. CER (customer edge router)
b. MPLS label
c. LER (label edge router)
d. LSR (label switching router)
e. FEC (Forwarding Equivalence Class)
f. LSP (label-switched path)
g. LDP (Label Distribution Protocol)
h. LIB (Label Information Base)

1. __ Employed by the entry label edge router to notify the downstream routers that a certain pathway has been allocated and that they should place labels in messages passing through along pathway.
2. __ A header placed in the address field between the layer 2 and layer 3 addresses that is used by the MPLS routers in the network to switch/route the frames and packets.
3. __ The router at the edge of a client's network that submits both requests for MPLS paths (and labels) and submits frames and packets for subsequent labeling and forwarding over the MPLS network.
4. __ The router at the entry point to MPLS network. This router receives customer request messages and applies an appropriate label to indicate the path that provided the transmission quality.
5. __ The base of information about a set of labels defining a chosen path that is forwarded to the LSRs along the chosen path.
6. __ Identifies a particular traffic flow along a path through the network. That preassigned path is designated a label-switched path (LSP) through the network, and each class is assigned to a particular LSP path.
7. __ A router within the MPLS network that switches/routes a message on the basis of the MPLS label that has been inserted between the level 2 and level 3 addresses.
8. __ A chain of labels that, in sequence, are applied to the message as it is forwarded from one router to the next along the designated path associated with the required bandwidth and service quality.

Short Essay Questions

1. What is Multiprotocol Label Switching?
2. Explain the similarities between ATM and MPLS.
3. Why is MPLS sometimes referred to as a layer 2.5 protocol?

Chapter 7

Local Area Network Technology

In the previous chapters, we looked at the use of circuit-switched networks. In circuit-switched networks, a dedicated path is carved out of a larger network to form a communications path. This path, or circuit, is set up by signaling, created when the communication is needed and terminated when the communication is ended. During this time, the path is exclusively the user's. A telephone call is a good example of the larger network referred to earlier. The public switched telephone network (PSTN) allocates an exclusive path for your call. When the caller hangs up the phone, the circuit is terminated and put back in the pool for the next caller to use.

The circuit-switched network is not only very effective but also very inefficient. During your phone call, the circuit is dedicated for your use even when no one is speaking. This waste of resources is similar to allocating a lane on the highway for your use with no one else allowed to use the lane until you have reached your destination.

An alternative to circuit switching is packet switching, a technology in which the path, or circuit, is shared by many users at once (similar to today's highway system). Packet switching allows an extremely efficient use of resources and, if properly implemented, is an effective mechanism for communication. Each packet, similar to a car on a highway, carries information that is destined for different locations.

A local area network (LAN) is an excellent example of packet-switching technology. A LAN is a privately owned network that provides communication to a local environment (typically less than 2 km). The network can support interfloor, interbuilding, and even intercampus communication, and can be used to connect local devices to a larger network. Another characteristic of the LAN is the packet-switching, or shared, environment. At any one point in time, different packets are present, each "owned" by different users and destined for different locations.

Business and Human Factors

Whether in a social or professional setting, people need to communicate with one another. In a business situation, this is not only a human need but also a business imperative. For work to be accomplished, people need to work together to solve problems and create resources for the company. LANs perform two essential functions: they (1) allow for information sharing and (2) provide resource sharing. These benefit the organization by allowing for improved decision making and, therefore, increased competitiveness.

During the 1980s, when personal computers (PCs), in significant numbers, were implemented in the working world, islands of information were created that were connected only to a human operator. Increased data and information were generated, but without an efficient and effective means to share the information, they were usually lost within the confines of a department. In the 1990s, the LAN connected these islands of information together, thus creating bridges (network) between users. The most common information-sharing applications are electronic mail, calendaring, and file transfer.

Another popular use of LANs is resource sharing. Essentially, many people share a single device. For example, each user with a PC may need to print files every so often. A printer can be obtained for each person, or a common printer (resource) can be shared among departmental personnel. The same concept of sharing a printer can be applied to many hardware devices (scanner, fax machine, etc.) as well as software. Rather than install a copy of a word processing package on each departmental PC, a single copy can be placed on a centralized resource, a server, to be used by many people. Resource sharing, although a bit more complicated in terms of implementation and delivery, leads to centralized management and support, decreases cost, and increases effectiveness of resources.

The use of LANs within an office environment has become a common and crucial element for information sharing and resource sharing, thus resulting in increased effectiveness at lower cost. LANs are also becoming prevalent in the home and in small remote offices. Often lumped into one category, small office/home office (SOHO) has the same purpose as office LANs, but it is smaller in size and lower in cost. For example, each of the PCs in your home can share a common printer (located, of course, in the kitchen), and your family can keep up a common calendar (sharing information that all can view and modify as needed). In a small remote office where personnel come in only when they need a connection to the corporate office, a LAN allows access to corporate information (such as pricing, vacation accrual, etc.) and to shared printer and fax services.

Costs of a LAN

The benefits of the LAN are obvious to those implementing it, but the costs are less obvious. When considering the cost of a LAN (or any type of network), the total cost of ownership (TCO) should be considered and factored in. Two broad categories of TCO are as follows:

- Initial one-time costs
- Ongoing costs

Initial Costs

The majority of the initial costs of a LAN include the hardware and software required to create the network. These include the following:

- End-user devices (PCs, etc.)
- Wiring
  - Cable costs
  - Installation costs
  - Termination costs
- Network equipment
  - Hubs, switches, and routers
- Network operating system (NOS)
- Shared devices
  - Servers
  - Printers
  - Fax machines

Depending on the numbers of users, the complexity of the applications, the distances, and the corporate standards in place, the costs can range from thousands to hundreds of thousands of dollars. One particular item in the list that should not be taken lightly is wiring installation. This element includes running the wires from each computer to centralized equipment (usually in a closet or basement). If the building is old (contains asbestos) or has hard obstacles (concrete floors), this one-time cost item can be extremely high. On the other hand, if the building is under construction and the walls are not yet in place, the installation is relatively easy and low in cost. These elements are discussed in detail in the technical section of this chapter.

Item	*Number*	*Cost*	*Total*	
Initial Costs				
Personal computers	10	$1,000	$10,000	
Network equipment				
Hubs	2	$500	$1,000	
Router	1	$1,500	$1,500	
Wiring	10 rooms	$400	$4,000	
Network operating system				
Shared devices				
Printer	1	$750	$750	
Fax	1	$300	$300	
Initial cost				$16,850
Ongoing Costs				
Personnel	¼ FTE	$7,000	$7,000	
Equipment maintenance		$2,000	$2,000	
Media access control address		$1,000	$1,000	
Outside line	$500/month	$6,000	$6,000	
Software license		$1,500	$1,500	
Ongoing cost				$17,500
Total cost, first year				**$34,350**

Exhibit 7.1 Total cost of ownership of a local area network.

Ongoing Costs

The other category of costs, the most-often-forgotten category, is the ongoing cost of ownership. To evaluate networks objectively, it is necessary to consider the subject of cost. LANs are an essential part of the way business is conducted these days and part of the cost of doing business. One should approach them with eyes wide open and be aware of them (Exhibit 7.1).

The items to consider are as follows:

- Personnel to maintain the network
- Equipment maintenance, upgrades, and repair contracts
- Moves, additions, and changes (MACs)
- Outside links
- Software licensing

Personnel

Just as we would never construct a building without factoring in the costs of cleaning and maintaining it, we should never consider constructing a network without also considering the costs of personnel to monitor, operate, and upgrade it. These personnel, usually highly trained and fairly expensive, provide for the daily and annual care and feeding of the network. The size of the network dictates the size of the support team (network administrators [NAs]). A small remote-office environment can survive using one person as the NA, designating one-fourth of his or her time spent on the network (after training); a 100-person office environment would require a full-time NA being assigned to this role and function.

Equipment Maintenance, Upgrades, and Repairs

All equipment is prone to breakdown, and replacement should be considered. Two strategies are prevalent: (1) obtain a maintenance contract that provides for replacement if something fails, or (2) take the risk and purchase a replacement in the event a unit fails. The criticality of your network and data usually dictates the strategy selected. If you want the network to be operational 24/7 with a 99.999% uptime, a maintenance contract is appropriate; if you can accept downtime of a few days, the second strategy is acceptable.

A positive note for consideration of the first strategy is support and upgrades. When getting a maintenance contract, vendors usually provide not only replacement of equipment in case of failure but also access to further support and upgrades for software. These two factors can have a critical effect on your network uptime.

Moves, Additions, and Changes

No matter how stable you think the network is, MACs occur. When a new office is partitioned from a large office, a new network connection needs to be added. Mr. Jones moves upstairs, so his connectivity needs to be moved and modified. Ms. Smith has been promoted, so her access to network resources has been changed. Each of these elements has associated personnel and a hardware cost. Depending on the fluidity of your organization, a sizable budget may need to be set aside for these MACs.

Outside Links

Usually, a LAN provides access to the Internet, to another corporation, or to headquarters. This connection of the LAN to the outside almost always carries ongoing costs with it. These costs can be significant ($4000 per year), depending on the speed, distance, and protocol used.

Software Licensing

The software used on the LAN is usually licensed (leased) for use on the network. Software includes computer operating systems (e.g., Windows), applications (e.g., Microsoft Word, Microsoft Excel), NOSs, and server operating systems. Software costs are sometimes a one-time investment, but if you want to upgrade the current version of the software, the costs are recurring. Depending on the number of users and the complexity of the software used, costs can be significant.

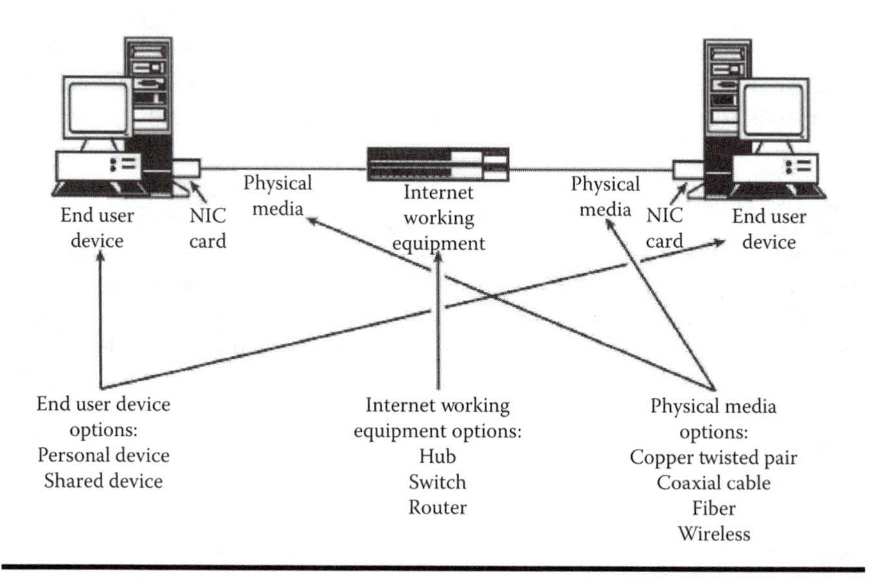

Exhibit 7.2 Components of a local area network.

Technical Factors

Working Parts of a LAN

A LAN has four essential components (Exhibit 7.2):

- End-user devices
- Physical media
- Networking equipment
- NOS

End-User Devices

The most visible parts of a LAN, the end-user devices (PCs, printers, smartphones, and servers), are what the users see and what they interface with, but they are not really part of the LAN. However, without these devices, the network would not be useful to the end users. End-user devices are the transmitters and receivers of information and the bridge between humans and the digital world.

Most end-user devices are personal in nature, meaning they are assigned to a single person. A desktop computer provided to an employee is a perfect example of a personal end-user device. However, there are also shared end-user devices, typically, printers, fax machines, and scanners that serve a department or a unit within an organization.

Name	Type	Data Rate (Mbps)	Distance (meters)	Often Used by	Cost ($ per foot)
Category 1	UTP	1	90	Modem	0.10
Category 2	UTP	4	90	Token Ring	0.05
Category 3	UTP/STP	10	100	10BaseT Ethernet	0.13
Category 4	UTP/STP	16	100	Token Ring-16	0.18
Category 5e	UTP/STP	100	200	100BaseT Ethernet	0.25
RG-58	Coaxial cable	10	185	10Base2 Ethernet	0.30
RG-8	Coaxial cable	10	500	10Base5 Ethernet	0.85
X3T95	Fiber	100	2,000	FDDI	1.00

Exhibit 7.3 Types of physical media.

Physical Media

The second network element is the physical media or wiring that connects the end-user devices to the network, allowing end users to communicate with them. Although not necessarily visible to the user, this is an essential component of the LAN. The physical media can come in many shapes and sizes (Exhibit 7.3), the selection of which is dependent on either what is currently existent in the building or the characteristics of the organization. The options include twisted-pair cable (i.e., category 3, 5e, 7, and 8) and fiber-optic cable. Twisted-pair cable is the most prevalent in today's LANs and the most effective way to accomplish building wiring. The signal tends to propagate down the wire efficiently, and because of the limited use of copper, it is a very cost-effective medium.

Coaxial cable is difficult to work with and higher in cost. Although still present in networking, it appears to be used less and less. Fiber-optic cable, although a bit more expensive to purchase and install, has the advantage of being almost totally immune to noise and external interference (lighting, motors, etc.) and, because of its higher speed and lower security risk, is being used more and more in the LAN environment.

Networking Equipment

The third component of a LAN is the networking equipment: switches, routers, and network interface cards (NICs). Often unseen by the user, hidden in closets and basements, this equipment forms the heart and brains of the network, sending information to the proper destination. Each of these devices performs the functions of getting the data from their source to their destination. Because they operate differently, a little discussion on each is appropriate.

Switches

Switches (Exhibit 7.4) provide a common physical access point and multiple ports for the physical connection of end-user devices. They come in 4-, 8-, and 16-port varieties, and selection is determined by the physical media used.

Exhibit 7.4 Picture of a switch. (Photograph compliments of Cisco System, Inc.)

Switches act differently with regard to receiving and transmitting packets (Exhibit 7.5). Switches act at layer 2 of the Open Systems Interconnection (OSI) model and transmit packets based on the media access control (MAC) address. A MAC address is a physical address permanently stamped in a device, in the NIC, and cannot easily be changed, much like your mailing address (unless you move).

The MAC address is learned by the switch through Address Resolution Protocol (ARP) requests. When the switch is turned on, it sends out ARPs to all devices connected, asking, "Who is there?" Each device dutifully responds with its MAC address, and the switch then creates a map of who lives at each port.

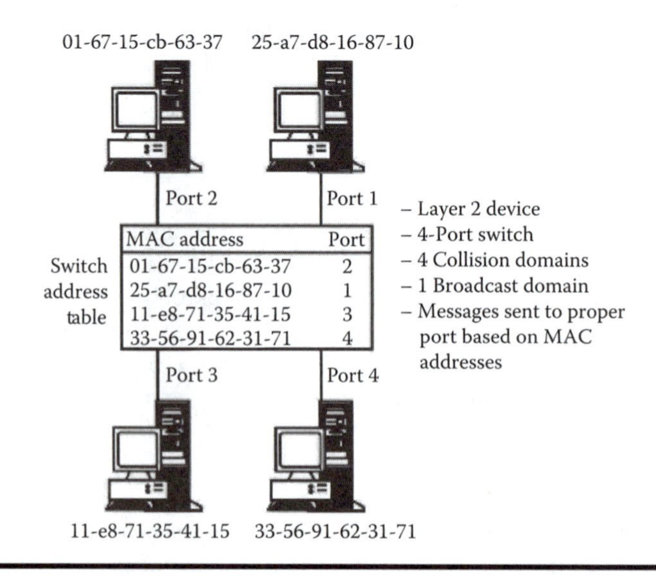

Exhibit 7.5 Switch operation.

On receiving a message, the switch looks to see where the packet is going, checks its map to see at which port the destination exists, and then repeats that packet out to the proper port. By sending the packet to the MAC address and not broadcasting to every port, network traffic is considerably reduced, and performance is enhanced. Switches can also be smart and translate between different physical media.

Routers

Although performing the same two core functions as switches (physical connectivity and message forwarding), routers are smarter and more effective in a network. Routers (Exhibit 7.6) operate at layer 3 of the OSI model (the network layer) and route packets to their proper destination address on the basis of the network address rather than the physical address. Recalling the street address analogy, a network address is a moniker given to a computer (based on protocol), similar to a family surname. Currently, the dominant network protocols, Transmission Control Protocol/Internet Protocol (TCP/IP) and Internetwork Packet Exchange (IPX), provide the PC a logical address (similar to a surname). Just as a family with a surname lives at a physical address, a TCP/IP address exists at a network MAC address.

Routers route packets to their destination based on the network address rather than the MAC address. What is the benefit? Every time you get a piece of mail addressed to the wrong person or a previous occupant at your house (physical address), you see the benefit: logical addresses are smarter and can find the correct location very quickly based on the network address.

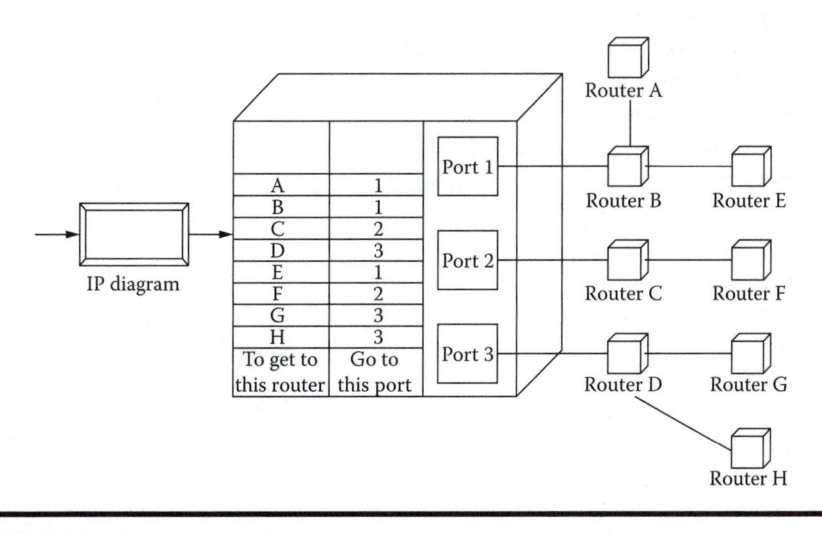

Exhibit 7.6 Router operation.

Routers learn addresses in ways similar to switches. Each router has a network address and knows where its neighbors are; therefore, it is much more capable of getting the packet to the right place. Needless to say, the benefits of this more effective routing of packets are speed and cost. Because a router has more to do, it takes longer to accomplish a task, and network performance could slow down. If used inappropriately, a router can slow a network down appreciably. Routers also tend to cost more than switches and are usually used at the edge of a LAN (where the LAN needs to connect to the outside world). The expression today is "switch first and route if you must," which means "use switches in a network first and routers only if you have to."

Although similar in appearance, switches and routers forward packets in vastly different ways. To confuse matters even more, there are many other offshoots of these: bridges, brouters, and layer 3 switches are but a few of the marketing labels put on devices that forward packets within LANs.

Network Interface Cards

NICs, pictured in Exhibit 7.7, lie between the end-user device and the network. They perform the translation and encoding necessary for a PC to talk through the network. PCs have different internal structures, called bus structures, which vary depending on the manufacturer and the type of PC. One must know the internal bus structure and find the correct NIC to fit it.

Additionally, the NIC must be compatible with the network it is connected to. Two criteria are important here: (1) physical media used and (2) speed. If the network uses coaxial cable, a physical connection must be made to that cable, and the NIC must be compatible with that connection. If you are using the more popular category 5e cable, the NIC must use that connector type. Speed is also important. If you are connecting to a category 5e cable that runs at 100 Mbps, the NIC must be capable of that connection speed.

Other important criteria should be considered in the purchase decision for the NIC, including management functions, protocols supported, reliability, and intended structure. On the basis of predominant use in the marketplace, the most popular NICs are PCI buses with a network connection for a category 5e cable at 10/100 Mbps. (These NICs can autonegotiate their speed for either 10 or 100 Mbps.) New computers are delivered with the NICs already installed, attesting to the increased popularity of LANs in the workplace and at home.

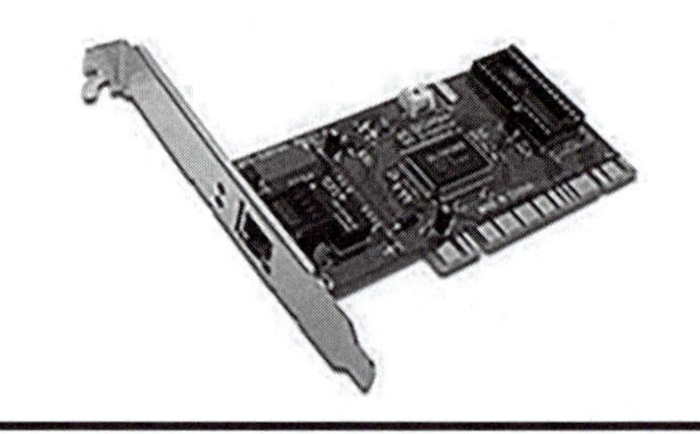

Exhibit 7.7 Picture of a network interface card.

Network Operating Systems

Software is needed to make the network equipment operate together. This software is called an NOS, which controls the order of communication between end-user devices (both personal and shared). Unlike a PC operating system, which controls an individual piece of equipment, the NOS controls communication between the devices connected to the network.

Common LAN Types

There are many types of LANs. This variety exists to meet the varying needs of the user base. Although there are many flavors, three bubble to the top as the most common: Ethernet, Fiber Distributed Data Interface (FIDDI), and Token Ring. However, one of these accounts for a vast majority of LAN traffic: Ethernet.

Ethernet (802.3)

Over 60% of the LANs worldwide use Ethernet. The Ethernet standard, developed by DEC, Xerox, and Intel, has become a de facto standard developed by the Institute of Electrical and Electronic Engineers (IEEE) 802.3 standard committee. The Ethernet LAN uses a logical bus topology. In a bus topology, all information is sent to a common core (the bus), where it is pulled off and listened to by all the computers. The computer to which the message is addressed "hears" its name and processes the message. Similar to calling out someone's name in a crowded hallway, the target computer (the person whose name you call) hears and processes the message. As for its advantage, the logical bus is efficient from a wide perspective and, not having a single point of failure, is less prone to failure. From a negative standpoint, the common logical bus may attract excessive traffic and may be a security concern.

From a physical standpoint, an Ethernet LAN is usually a star topology (Exhibit 7.8). All the connecting devices go into a switch, where they are internally connected to a bus. When conceptually thinking about networks, we think of them in their logical format; when we actually hook up the wires, we find, for convenience's sake, a physical star. Simply put, Ethernet topology is a logical bus and a physical star.

Several end-user devices share this same logical bus; so, it is important to control access to the bus. If two computers want to communicate at the same time, their packets collide on the network and become jumbled. To stop these collisions, an access control method must be used. Ethernet uses an access control method called Collision Sense Multiple Access/Collision Detection (CSMA/CD). With CSMA/CD, computers "listen" until they hear nothing (meaning the bus is clear), and then they put their packets on the bus (the CSMA part). If the bus is busy, the computer waits until the bus is clear before it transmits.

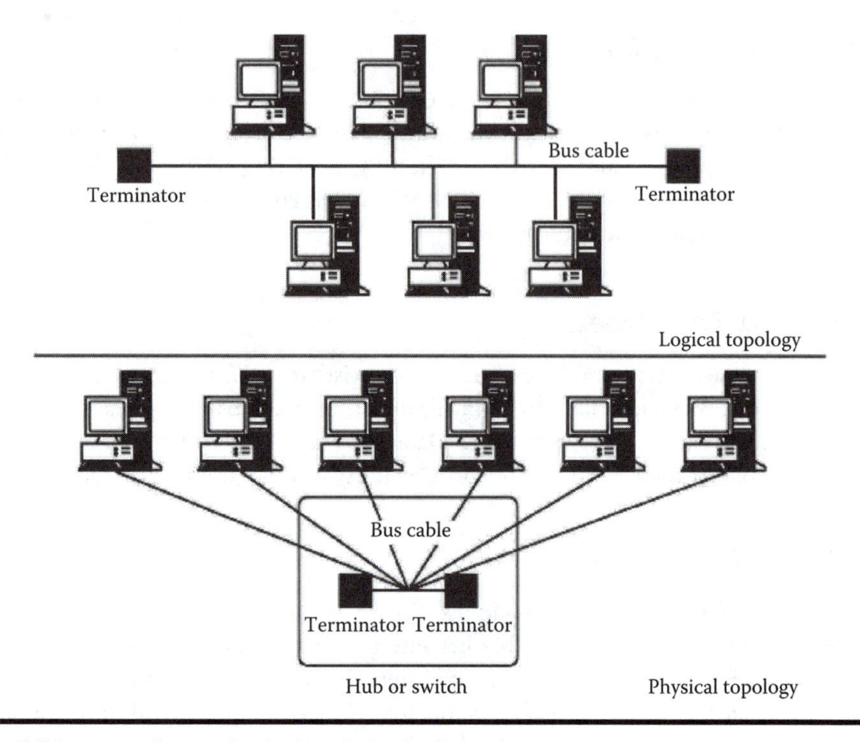

Exhibit 7.8 Ethernet logical and physical topology.

Suppose two computers want to communicate. They both listen to the bus and note that it is clear. Both put their packets on the bus at the same time. In this case, a collision occurs. This is where collision detection comes in. After putting the packets on the bus, the computer continues to listen, and if it hears a collision (packets other than its own on the bus), it immediately stops transmission. Both computers then wait a random amount of time before transmission begins again. Because the computers are waiting random periods of time, it is very unlikely that the two will start transmission again at the same time, causing more collisions.

Ethernet is a baseband type of transmission, meaning that only one signal is on the bus at a given time. The common physical media used with Ethernet is twisted-pair cable (category 5e), although, as noted before, coaxial cable and fiber-optic cable are also used.

Although originally specified at 10 Mbps, Ethernet is now commonly used at 100 Mbps/1 Gbps, and specifications for 10 Gbps have been completed. Generally, as the speed of the network increases, the distance limitation between the end-user devices and the network equipment (switch and router) decreases. This is often a critical factor in assuring communication to all people in a building; so, this must be assessed carefully.

Regulatory and Legal Factors

Use and Function of Standards

Standards—hard to live with them; impossible to live without them. Standards are sets of guidelines or rules that people and corporations agree to follow to make the world an easier place to live and work in. Imagine if tire sizes were not standardized and we had to go to the manufacturer of our car for new tires. Imagine if lamps used unique-sized light bulbs and only the lamp manufacturers carried the specific bulb.

Without standards, each company can do its own thing, creating products that it thinks are best for you. With standards, companies are often limited in their creativity and uniqueness, but as consumers, we get the following:

- Lower costs: Competition for the same market often provides lower cost.
- Flexibility: A network can have equipment from multiple vendors; therefore, users are not locked into a specific vendor.
- Compatibility of equipment.

Unless there is a unique and critical need, and if money is no object for your company, or if your company is on the cutting edge of its industry, you will be using standardized equipment, and rightfully so. Standards make the operation of a system and organization go much more smoothly and give you flexibility in your actions.

Process of Standardization

There are essentially two types of standards: de facto and formal. De facto standards are agreements that have no official or legal meaning. For example, the Windows operating system is a de facto standard that companies other than Microsoft write programs and applications for. Because of Microsoft's position, there is a plethora of programs available for our computer (which is great), and prices are dropping daily. Most corporations purchase computers with Microsoft products on them, strengthening the de facto standard. On the other hand, a formal standard is one developed by an official industry or government entity. The process of creating a formal standard is threefold:

- Specification: The entity identifies the problem to be addressed, the issue involved, and the terminology to be used.
- Alternative identification: The alternatives are identified, and the pros and cons of each are brought out.
- Acceptance: The group agrees on the best solution and garners support from leaders in the field.

A lot of fame and fortune can be obtained from this process and the resultant standard, so it is not without its problems and hidden agendas, which, along with the myriad of issues, complexities, and human considerations, often make the standard-setting process very lengthy. It often takes 3 to 5 years to start and complete a formal standard. With technology changing so rapidly, by the time a standard becomes operational, it may no longer be needed.

Standard-Setting Bodies

There are numerous standard-setting bodies; because of the nature of the field, most of them are international. The more visible ones are as follows:

- International Organization for Standardization (ISO): Because the abbreviation comes from the French version of the name, it is *ISO* and not *IOS*. Based in Geneva, Switzerland, and international in nature, ISO members come from the national standards organizations of each member country. Its task as part of the International Telecommunications Union (ITU, discussed next) is to make recommendations about data communications interfaces and telephones, as well as telegraph systems, on an international basis.
- ITU: This body is the standards-setting arm of the United Nations International Telecommunications Union. Formerly known as the Consultative Committee on International Telegraph and Telephone (CCITT), it has representatives in more than 160 countries. ITU formulates recommendations for use by the telephone and telegraph industry, common carriers, and hardware and software vendors.
- American National Standards Institute (ANSI): Consisting of over 900 vendors, it is the coordinating arm for the US system of standards. Its role is to coordinate the development of national standards and to interact with international standards bodies so that they are aligned with one another.
- Electronic Industries Association (EIA): EIA is a body that determines equipment and hardware standards. The common RS232 cable connection standard is a result of EIA. Its members come from telecommunications vendors.
- IEEE: A professional society based in the United States that focuses on LAN standards.

Legal Aspects

Although these bodies set standards, they do not make laws. In other words, the standards are not legally enforceable in any country. They are formal agreements that make the world an easier place to operate, but they do not have to be followed. Vendors follow them to sell their products; consumers buy standardized products because of the benefits.

Other Networks

An understanding of LANs provides a solid starting point in understanding other network technologies. In subsequent chapters, you will encounter a number of these networks, such as wide area networks (WANs), wireless local area networks (WLANs), storage area networks (SANs), virtual private networks (VPNs), and more.

One increasingly prominent network type is the metropolitan area network (MAN). A MAN is generally a network larger than a LAN but not as large as a WAN, comprised of separate buildings within a single city. Interbuilding communication has historically been limited by telephone service offerings, which were themselves limited by the underlying telephone facilities. There are two main forms of MANs: wireless and Metro Ethernet MANs. Wireless metropolitan area networks (WMANs) will be discussed in Chapter 9.

Metro Ethernet MANs take advantage of the high concentration of Ethernet infrastructure (copper and fiber) used to connect carrier sites and central offices (COs) in a city to link dispersed buildings and organizations. Most national carrier networks are already constructed in a packet-like network format in which digitized telephone calls are segmented into small pieces, with each piece placed into small, fixed-sized Asynchronous Transfer Mode (ATM) cells (ATM fixed-length packets). These voice-carrying cells are then placed into large synchronous optical network (SONET) frames for bulk transport across the national network. MPLS service may offer label switching as an overlay switching component and possibly eventual replacement of the SONET component. All traffic is intermingled and transmitted together as a bundled unit over the carrier network.

Summary

In this chapter, we explored the packet-switching technology of networks, specifically LANs. We first looked at why LANs are used in business and what benefit they provide to business and private users. Next, we explored the costs of a LAN and defined the TCO.

We then delved into the technical factors associated with a LAN. We looked in depth at the four parts of a LAN (end-user devices, physical media, networking equipment, and NOS) and at the various flavors of LANs. Both the logical and physical layout of Ethernet LANs were discussed.

Finally, we discussed the reasons that standards are important, and we reviewed the process of setting standards. We also listed the various standard-making bodies that exist in the world today.

Exhibit 7.9 describes a case study that applies to the technology discussed in this chapter.

Eighteen vendors of 10G Ethernet products contributed their wares for a demonstration of the technology at NetWorld+Interop 2001 in Atlanta in an effort to show the network industry that prestandard 10G technology can successfully communicate.

The product test fest was hosted by the 10 Gigabit Ethernet Alliance (10GEA) industry group and was billed as more of a general product demonstration than a proving ground to show whose products worked and whose did not, so results on vendor-to-vendor interoperability were not released. Even so, demo organizers say the showing of 10G products proves to enterprise and carrier users that 10G bit/sec Ethernet is ready for deployment and that the technology's hefty price tag could drop soon, thanks to advancements being made in optical interface and 10G chip technologies.

Major Gigabit Ethernet switch vendors that participated in the demo included Avaya, Cisco, Extreme Networks, Foundry Networks, and Nortel. All of these switch vendors have announced 10G products that are supposed to ship by year-end, except for Extreme, which has not yet specified a shipping date or price for its product. Component and testing equipment makers represented at the demonstration included Agilent Technologies, Broadcom, and PMC-Sierra.

"The focus was getting the vendors to play together and to get traffic flowing," over 10G connections, says Bob Grow, chair of the 10GEA and an Intel engineer.

Instead of just connecting all the vendors' 10G switches together in one large 10G bit/sec mesh, the switches were set up in a mixture of topologies, with some switches aggregating multiple gigabit connections and a few connecting to each other over 10G links.

Five of the seven vendors' switches in the demonstration were able to talk to each other over 10G links, according to Grow.

[a] Phil Hochmuth, *Network World*, 9/17/01.

Exhibit 7.9 Case study: 10 gigabit Ethernet put to the test at NetWorld+Interopa.

In the following chapter, we will delve further into networks, exploring the language and routing protocols that get your message to the intended receiver.

Review Questions

True/False

1. LANs are inefficient for resource sharing.
   a. True
   b. False
2. The total cost of ownership should take into account initial costs and ongoing costs.
   a. True
   b. False
3. If you are in a stable organization, you should not have to make moves, additions, or changes to your network configuration.
   a. True
   b. False
4. Coaxial cable is no longer used as a networking medium.
   a. True
   b. False

5. NICs are the connections between the end-user device and the network.
   a. True
   b. False
6. Ethernet uses a logical bus and a physical star topology.
   a. True
   b. False
7. Ethernet is identified as the IEEE 802.3 standard.
   a. True
   b. False

Multiple Choice

1. MAC stands for:
   a. Media access control
   b. Moves, additions, and changes
   c. All of the above
   d. None of the above
2. Ethernet usually uses which of the following topologies?
   a. Physical bus
   b. Logical bus
   c. Physical ring
   d. Logical ring
3. Which of the following technologies uses CSMA/CD?
   a. Ethernet
   b. Frame Relay
   c. All of the above
   d. None of the above
4. What is not a characteristic of a LAN?
   a. Uses circuit-switching technology
   b. Is a privately owned network
   c. Provides communication to a local environment
   d. Is used for resource sharing
5. A MAC address is:
   a. Configured by the network administrator
   b. Found on the computer's network interface card
   c. Assigned by the Internet provider
   d. Found on the computer's hard drive
6. Which of the following is not an ongoing cost of network ownership?
   a. Installation cost
   b. Personnel to maintain the network
   c. Equipment maintenance and upgrades
   d. Software licensing

7. Which of the following is not a consideration in the selection of the NIC?
   a. The physical media being used
   b. The network speed
   c. The type of network it is being connected to
   d. The type of computer being used

Short Essay Questions

1. What is CSMA/CD? How does it work, and what technology does it work with?
2. Name and describe the four components of a LAN.
3. Describe packet switching. Compare and contrast it with circuit switching.
4. Name and describe two considerations for network costs.

Chapter 8

The Language of the Internet: Transmission Control Protocol/Internet Protocol (TCP/IP)

This chapter explores the language and functions used in data communication and the Internet. Data communication, specifically over the Internet, has a language all its own. This language of the Internet, Transmission Control Protocol/Internet Protocol (TCP/IP), was developed as a universal language that would serve as a universal standard when possible to avoid all the pitfalls of picking any one proprietary system.

There are many languages for data communication. TCP/IP is certainly not the only one, or the best one, but it is the language of the Internet (a very popular tool in today's culture). We will look at this language in depth in this chapter and will introduce you to a few other languages that are used. With the information from the previous chapters, you will develop a better and deeper understanding of the way data communications operate within the business environment.

Business and Human Factors

Humans have a need to communicate, and we often meet this need through technology. By providing mechanisms, standards, and connections, we allow machines and humans to communicate.

One of the fastest-growing mechanisms for communication today is electronic mail, or e-mail, which allows fast and cheap communication between people. The biggest advantage of e-mail is its asynchronous nature, allowing communication that is not real time. We all have had the need to contact someone, and we pick up the phone and dial the number, only to be met either by a busy signal or no answer at all. E-mail allows us to send a message to a person and have that message responded to when the recipient is available.

E-mail is certainly not the panacea of communication and does not serve as the best mode of communication for certain needs, like urgent messages. It is just another tool that can, if used effectively, round out our repertoire of communication methods and meet our need to get and give information.

The growth and popularity of the Internet has been due to many factors, but e-mail has certainly been the greatest contributor to its growth and the Internet's most popular application. Whether you measure an application's popularity by the availability or by the number of users, e-mail is the killer application of the Internet.

One of the key reasons for the success of e-mail is the ability to send mail to almost anyone anywhere in the world for a minimal amount of money. Added to that is the ease of putting together an e-mail and the quick travel time. It is now common upon meeting someone to ask for his or her telephone number and e-mail address.

What once was a medium for "techno-savvy" people only is now as important as the telephone; in fact, e-mail may be even more important than the telephone. In today's business environment of international corporations, where time zones become a severe hindrance, the asynchronous nature of e-mail allows people to communicate at their convenience. Recently, the American Management Association found that more than 57% of business executives in America rely on e-mail for their work.

Corporations also find that e-mail is cheaper than local mail, and companies want to leverage their investment in Internet technologies. When companies want to extend their customer support base and improve management communication and interaction with customers and suppliers, e-mail seems to fit the bill well.

From a human perspective, e-mail use has also changed. Some intriguing characteristics are the following:

- The average size of an e-mail message is increasing, with the growth attributed to attachments.
- There are approximately 440 million e-mail addresses worldwide.
- By the end of 2001, the average number of messages received was 35 per person per day.
- In 1999, 200 billion pieces of mail were delivered by the US postal system. E-mail has outpaced regular mail volume, with 560 billion messages being delivered per year.

- An estimated 20% of all e-mail received is commercially driven. This is probably less than the junk mail we receive in the postal mailbox.
- Despite challenges from instant message services and virtual workspaces, e-mail is still the clear winner in Internet use.

E-mail is clearly not an ideal solution for all situations. It is not effective for sensitive conversations or for information that is needed instantly. However, e-mail is another helpful tool when passing information is key and where time differences and working styles are different.

Technical Factors

Message Casting

Messages sent on a network are categorized by the number of recipients they are directed to. The three basic categories of messages are unicast, multicast, and broadcast:

Unicasting: With unicasting, the sender transmits the message in the form of packets addressed to a single destination (Exhibit 8.1). A unicast message is intended to go from one computer to another computer. It is similar to a person in a crowded room calling out a particular name (which tells an intended party who the message is addressed to).

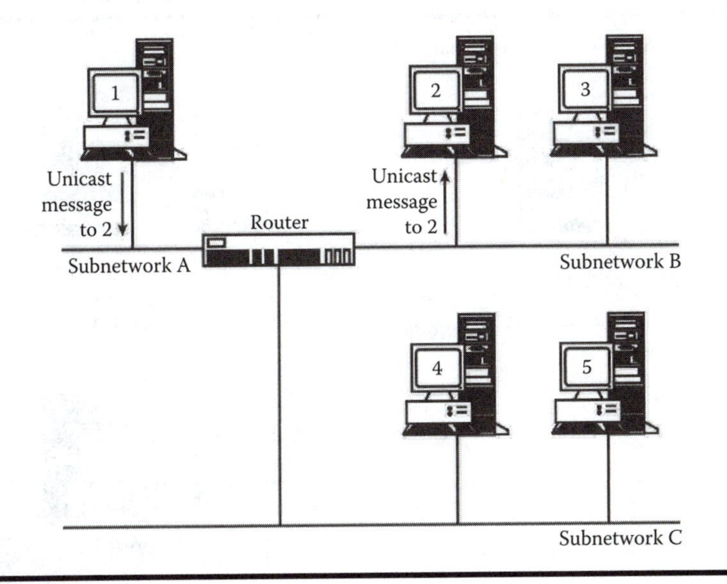

Exhibit 8.1 Unicast transmission.

Multicasting: With multicasting, a message is transmitted to people in a distinct group or category (Exhibit 8.2). If you called out for all females to listen to you, each of them, knowing their group or classification membership, would then listen to the message.

Broadcasting: With broadcasting (Exhibit 8.3), a message is transmitted to an entire group bound together by a physical restriction. If you called out to an

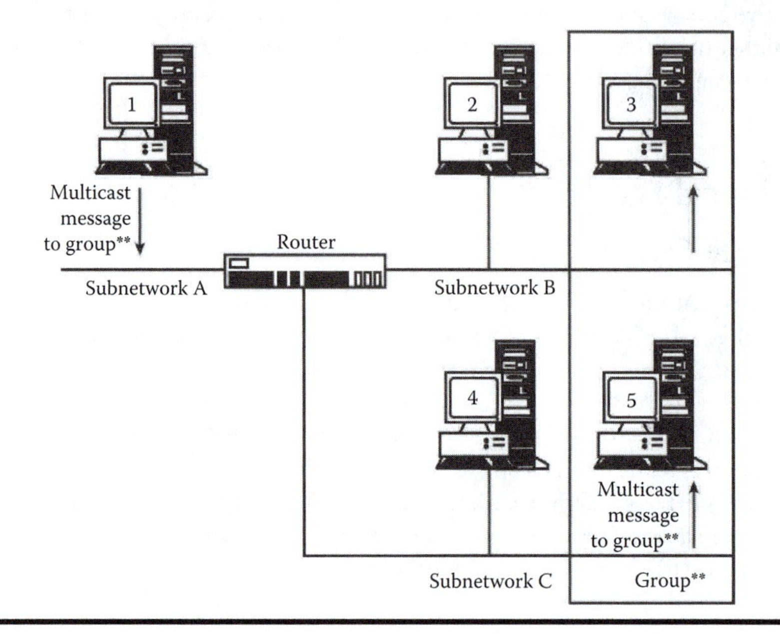

Exhibit 8.2 Multicast transmission.

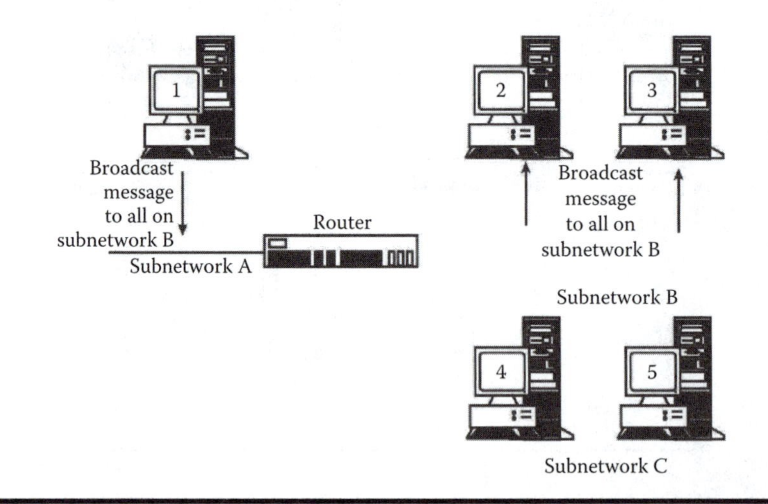

Exhibit 8.3 Broadcast transmission.

entire room, "Everybody listen," everyone in the room would respectfully do so. Broadcasting on a network means that every computer or server on the network or network segment receives and processes the message.

For each of these conditions to exist, the intended sender and receiver must know certain information about each other. For example, in our analogy, everyone must know their name, their gender, and what room they are in. The sender also must know the same information for communication to occur. Obviously, some mechanism or standard must be agreed upon so that everybody can communicate effectively. These standards, or agreements in method, must exist and are part of layer 3 of the Open System Interconnection (OSI) model, the network layer. At layer 3, schemes for message addressing all network devices are agreed upon and subsequently used for communication to happen effectively.

Message Addressing

For effective message transmission, whether unicasting, multicasting, or broadcasting, each device on the network must have a distinct and unique global address, and the address must be understood by each member on the network.

TCP/IP is one of the most predominant networking languages (routed protocols); thus, it has a unique naming and addressing scheme. Each routed protocol has a different naming and addressing scheme, but we will look only at TCP/IP.

In the case of TCP/IP, addresses are broken down into two separate parts: network ID and host ID. The network ID is similar to the city and state in a mailing address. This information tells the post office the general location of your street but it is not exact enough to pinpoint your house. The host ID portion of a TCP/IP address gives the specific location of your computer on your network (similar to your house number on your street) and therefore assures a unique address.

IP addresses are written in numerical form, similar to a telephone number, but in four distinct parts. Each of the parts is designated as part of either the network ID or the host ID. Each of these parts is called an octet and is separated by a period or dot: Part1.Part2.Part3.Part4.

In reality, each part is a number between 0 and 255. So, an address can be as follows: 127.52.16.4.

Now the question arises: which of the parts or octets are part of the network ID, and which are part of the host ID? Our example of the street address shows some similarities:

Street address	**TCP/IP address**
John Doe	127.52.16.4 123
Main Street	
Whitestone, New York	

In a street address, the city and state are always on the bottom line. This information can be thought of as the network address. The network address, and the city and state, provide the general location but not the exact address of your house. The street address and the name (lines 1 and 2) provide the exact location. The host portion of the IP address provides this precise information; but with a TCP/IP address,

Class	*Number of addresses in Hot Portion*	*Number range*	*Breakdown*
A	16,777,214	1.0.0.0–127.0.0.0	N.H.H.H
B	65,534	128.1.0.0–191.259.225.225	N.N.H.H
C	254	192.0.0.0–223.255.255.755	N.N.N.H

Note: N = network, H = host.

Exhibit 8.4 Classes of TCP/IP.

State University is a midsize university in the Midwest. It has approximately 20,000 students distributed over seven colleges. When State University applied to the InterNIC (an organization that is charged with registering domain names and IP addresses as well as distributing information about the Internet), it was given a class B license based on its current size and projection of growth: 136.15.0.0. With a B license, the first two octets define the network ID portion of the address (136.15). These octets define and differentiate State University from all other colleges, commercial firms, and organizations in the world. The last two octets are left to State University to use as host ID addresses.

State University is an organized and methodical place; it has decided to separate the host addresses for each college and each computer. Thus, the addresses for each computer look like this:

College of Arts and Sciences
136.15.1.1.—Joe Dean's computer
136.15.1.2.—Sally Secretary's computer
...
136.15.1.254.—Tom Teacher's computer
College of Fine Arts
136.15.2.1.—Frank Dean's computer
136.15.2.2.—Sam Receptionist's computer
...
136.15.2.254.—Kristen Teacher's computer

The internetworking equipment reads each address and determines which class of license it is (from the number in the first octet), which network it is, where the network is, and finally, which computer of that network the message goes to.

Exhibit 8.5 Case study: State University.

the octets vary by the class of address. There are three common classes of TCP/IP addresses, as shown in Exhibit 8.4.

Class A licenses are for large organizations and provide large numbers of octets or host IDs for local use. Class B licenses are for medium-sized organizations and allow for two of the octets for host IDs. Class C licenses are used for small organizations and allow for only the last octet to be used for host addresses. Referring to the breakdown portion of Exhibit 8.4, we see the number of octet labels for either the network (N) or the host (H) portion.

Depending on the size and intentions of an organization, a particular class of license is allocated; the license then defines which of the four octets are the network ID portion and which are the host ID portion of the address (Exhibit 8.5).

Network Routing

To understand how addresses work, you must first understand how the messages are routed through the network to get to their destination. Routing is the process for delivery that packets undergo from the sender's computer to the receiver's computer through the network.

For many of the LAN technologies and topologies we have looked at, the routing is simple. The message goes to all devices on an Ethernet bus or to each individual device in a Token Ring, and the receiver hears its address and processes the message. However, the Internet is complex, and routing is a difficult mechanism that is critical to getting the message to the right place (Exhibit 8.6). There are essentially two categories of routing, centralized and decentralized.

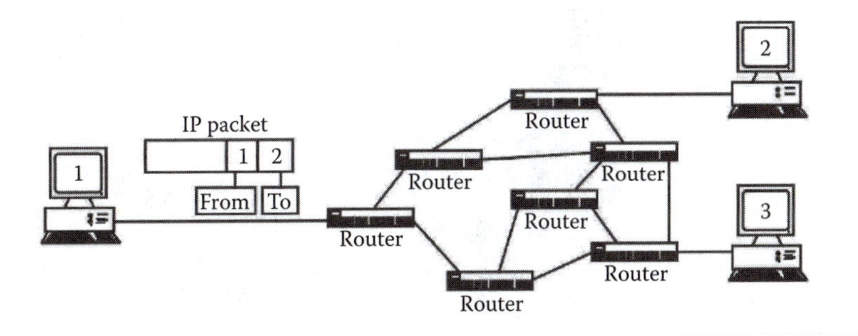

Exhibit 8.6 Message routing.

Centralized Routing

In centralized routing (Exhibit 8.7), all decisions about the path a packet is going to take are made by a central device. In a star topology, all packets are routed to the central device, where decisions are then made on how to forward the packet to its destination. The central device, a telephone switch or a router, learns which devices are connected to it and then creates a table to keep track of the address and location of each device. The table is used to look up each connected device to determine the path it must take. Even in a complex topology, a central device can maintain an effective routing table and route the packet to its intended destination.

The primary advantage of centralized routing is its network efficiency. One routing table that keeps track of the whole network is maintained, and thus, computer resources and precious network capacity are managed by concentrating all the work on one device. This primary advantage is also its biggest disadvantage. If the central device fails, the network fails. Without a brain to tell the packets where to go, they will wander aimlessly around the network until they finally die. Another disadvantage of centralized routing is scalability. As the network grows and more devices are connected to it, the central routing table can become very large and unmanageable. These disadvantages and the need for networks that always work have led to the advent of decentralized routing.

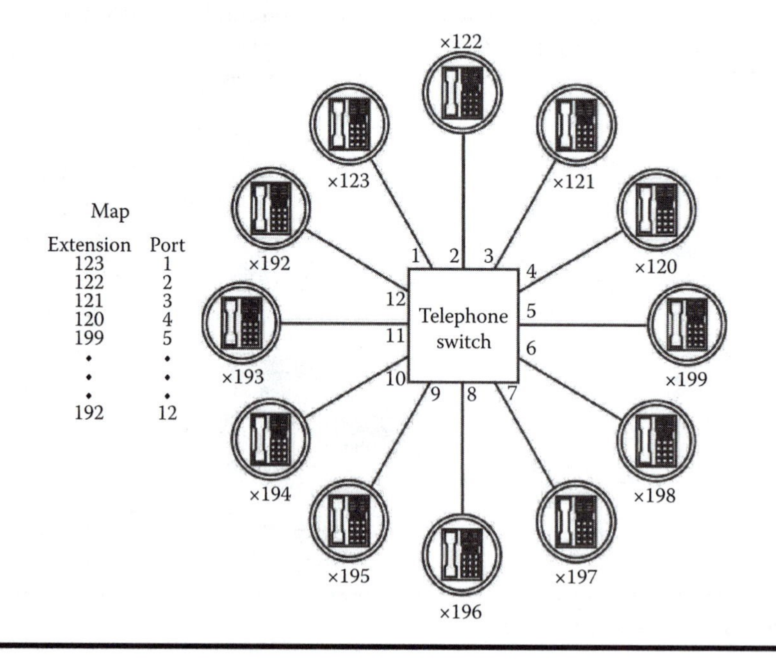

Exhibit 8.7 Centralized routing.

Decentralized Routing

In a decentralized routing scheme, no one device maintains the routing table (Exhibit 8.8), but many devices maintain parts of it. This may seem chaotic (too many cooks spoil the broth) and inefficient (many machines performing the same function), but in reality, it works very effectively.

The key to this effective operation is a set of rules called *routing protocols*. Routing protocols are rules that govern how each device works within the whole network and how the whole network operates as a system. For example, in an office, one person is responsible for sorting the mail, another is responsible for delivering the mail, and another is responsible for filing the mail. Overall, they are each part of a large system that ensures that the mail system operates effectively.

There is no best routing protocol, and there are various flavors of rules to choose from. Each protocol is a set of procedures that makes the system work if all the members in the network follow the rules.

These routing protocols have been accepted by the industry and have been standardized in their operation so they can be used across different vendor platforms. Basically, the routing protocol is selected by the type and size of the network (Exhibit 8.9).

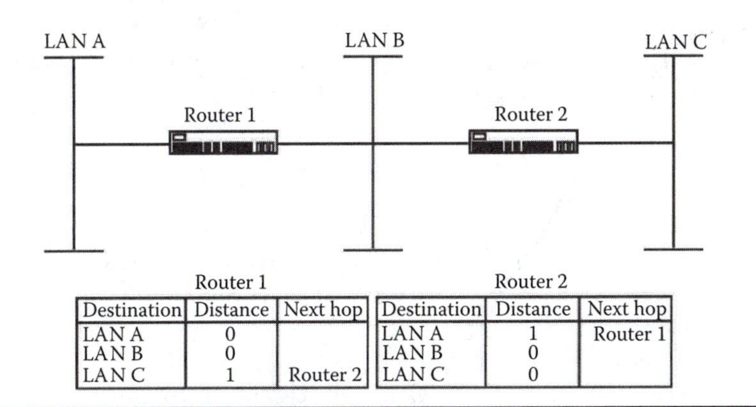

Router 1

Destination	Distance	Next hop
LAN A	0	
LAN B	0	
LAN C	1	Router 2

Router 2

Destination	Distance	Next hop
LAN A	1	Router 1
LAN B	0	
LAN C	0	

Exhibit 8.8 Decentralized routing.

	LANs	*WANs*
Small	RIP	OSPF
	IGRP (proprietary)	
Large	OSPF	OSPF
	EIGRP (proprietary)	BGP-4

Exhibit 8.9 Routing protocol.

Some of the routing protocols are proprietary, which means they work only with a specific vendor's equipment. If you have this specific vendor's equipment, it is necessary to select the proprietary routing protocol. If you have a multivendor environment, or use a different vendor's equipment, you must select other routing protocols, for example, Routing Information Protocol (RIP), Open Shortest Path First (OSPF), or Border Gateway Protocol (BGP). These routing protocols are standardized and run across different platforms and systems.

Language of the Network: Routed Protocols

We have described how packets are routed using routing protocols along a path to their correct destinations but not how these packets are arranged (routed protocols). The packets must be arranged in a specific form according to a set of rules to be understood. Similar to language, these routed protocols have rules; that is, there must be a verb in each sentence, words must match in tense, etc.

These network languages operate predominantly at the network layer (layer 3) of the OSI model and form the information we wish to communicate. The languages have become standardized, so they can be understood by many different vendors and computers. Although there is a plethora of standardized and proprietary languages, we will look at three of the most common: TCP/IP, Internetwork Packet Exchange (IPX)/Sequenced Packet Exchange (SPX), and Systems Network Architecture (SNA). Although each language operates differently, there are many similarities between them, and it is encouraged to understand these similarities.

TCP/IP

The language of the Internet, TCP/IP, is the most common and the oldest standardized routed protocol. Because it can use a variety of data-link protocols (Ethernet, Token Ring, ATM, frame relay), it is a very popular and effective protocol. Coupled with its ability to transmit messages without errors (or at least to correct its errors), it is the language most networks and equipment speak.

As you may have guessed, TCP/IP is made up of two separate parts: TCP and IP. IP was developed in the 1970s as a way of packetizing and sending messages across what was called the Advanced Research Project Agency (ARPA) network. Operating at the network layer of the OSI model, the main function of IP is the addressing and routing of packets. Because IP functions on all the networking devices, the network can route the packet using the IP addresses and IP routing protocols (RIP, OSPF, etc.) This function provides the method for getting a packet or a message from the sender to the receiver.

Exhibit 8.10 shows the detailed breakdown of an IP packet (IP version 4 [IPv4]). Each of the parts (think of them as verbs, nouns, adjectives, etc.) serves a specific purpose:

Version	Type of service	Total length	Flags	Hop limit	Protocol	Source address	Destination address	Data

Exhibit 8.10 IP packet structure IPv4.

- Version number: Shows the version of the IP language. Version 4 is the most common version used today.
- Type of service: A field that can be used to identify the type of packet (voice, data, or video message) and allow for prioritization or quality of service (QoS, discussed later).
- Hop limit: To stop lost packets from forever traversing across a network, a limit is imposed on the number of hops or transmissions from one router to another. This field essentially controls the life of the packet in the network.
- Source address: The IP address from which the message originated.
- Destination address: The IP address to which the packet is destined.
- User data: This is where your message (an e-mail, a word processing file, or a phone call) is actually placed. The other parts of IP are all business overhead but nevertheless essential. The user data field can vary from 8 to 1428 bytes in length.

The other part of TCP/IP is TCP. TCP performs a few essential functions and couples well with the IP portion of TCP/IP. Its main functions are as follows:

- Packs and unpacks the data: Sometimes, a message cannot fit into one packet, so it must be broken down into many packets and reassembled at the other end. TCP performs this function but must be operating at both ends of the network to break up and reassemble the data.
- Ensures reliability: IP by itself is similar to the postal system; you send the message and pray that it reaches its intended recipient. Combining TCP and IP is similar to sending mail "guaranteed," return receipt requested. TCP waits to hear from the receiver that it received the message. If the message is lost or corrupted during its travels, TCP resends the packets. A typical TCP packet is shown in Exhibit 8.11.

A TCP packet is a bit different from an IP packet. You can see the fields that are provided for successful packing and unpacking (sequence #) and to ensure reliability (ACK [acknowledgment] #). The sequence number lists the part of the total

Source port	Destination port	Sequence #	ACK #	Length	Flow control	Options	Data

Exhibit 8.11 TCP packet.

this packet is from (e.g., 2 out of 3). The acknowledgment section is received after the receiver confirms its receipt.

IPX/SPX

IPX was developed by Xerox in the late 1970s and became a proprietary packet protocol used with Novell Netware. Because Novell has a fairly large market share, the language is common on networks and probably will not disappear in our lifetime. Novell has recently introduced its new network product that uses TCP/IP instead of IPX/SPX for encapsulation.

As the name implies, IPX/SPX is made up of two essential pieces (Exhibit 8.12). The SPX portion behaves in much the same manner as TCP and has the same function to perform (ensuring reliability, and packing and unpacking the data). The IPX portion of the protocol behaves very much like the IP portion of TCP/IP. The major difference is in the addressing: as part of the address, IPX incorporates the media access control (MAC) address of the devices; thus, the address of an IPX device has both the layer 2 and layer 3 information required for communication to occur.

SNA

IBM developed a proprietary network protocol called SNA in the late 1970s. The protocol is very effective on IBM machines. SNA is not just a routing protocol or a routed protocol but a full telecommunications architecture.

A bit of history might help here. IBM was one of the pioneers in computers. Developing the mainframe and personal computers, IBM rose to prominence and even dominance in the field. Understanding that people need to communicate over distances, and recognizing the lack of a routing language, IBM developed SNA as a protocol for networking.

In the late 1980s, a philosophical shift away from proprietary architecture toward an open system and mixed vendor networks took hold in both computing and networking. TCP/IP was embraced as the panacea; anything proprietary was discounted. As a company, IBM suffered many losses in the early 1990s. Many reputable sources predicted that its mainframe and architecture would wither away by the 21st century.

Checksum	*Lensil*	*Control*	*Destination Address*	*Destination Network Address*	*Destination Socket*	*Source Address*	*Source Network Address*	*Source Socket*	*User Data*
2 bytes	2 bytes	1 byte	6 bytes	4 bytes	2 bytes	6 bytes	4 bytes	2 bytes	Varies

Exhibit 8.12 IPX packet structure.

Both IBM and its mainframe are still around, and in fact, both are doing amazingly well. Despite many projections to the contrary, there is still a need for mainframe computing power and reliability. Many of its architectures still exist, and SNA is certainly one that will be encountered in the workplace. It is necessary to have a minimal understanding of SNA.

In SNA parlance, the end-user terminals are designated logical units (LUs). These LUs are end points to the SNA network, and their code is programmed into the device. LUs talk to one another via sessions. The controlling network software, upon a request from an LU, sets up a session with another LU (another terminal, a mainframe computer, or a microcomputer). Before the session actually begins, rules (amount of data to be sent, frame size, etc.) are specified by the two LU devices. Each LU having a coded number (similar to a MAC address) is also assigned a network name that allows LUs to talk to another name regardless of its location (The network makes the translation between name and coded number.) As the session progresses, the packets move along in a Synchronous Data Link Control (SDLC) frame. The session is ended when one of the LUs sends a deactivation request (Exhibit 8.13).

Besides LUs as SNA-addressable devices, there are also physical units (PUs) and system service control ports (SSCPs) (Exhibit 8.14). PUs can be the end terminals and microcomputers, and they also can be a terminal control or a computer front end. These PUs have an intelligent role to play in the large-scale SNA network. The SSCP, a set of SNA components, manages the entire SNA network or a part of the SNA network (called a *domain*).

Another part of the SNA world, besides the network-addressable units discussed previously, is the path control network. The path control network provides

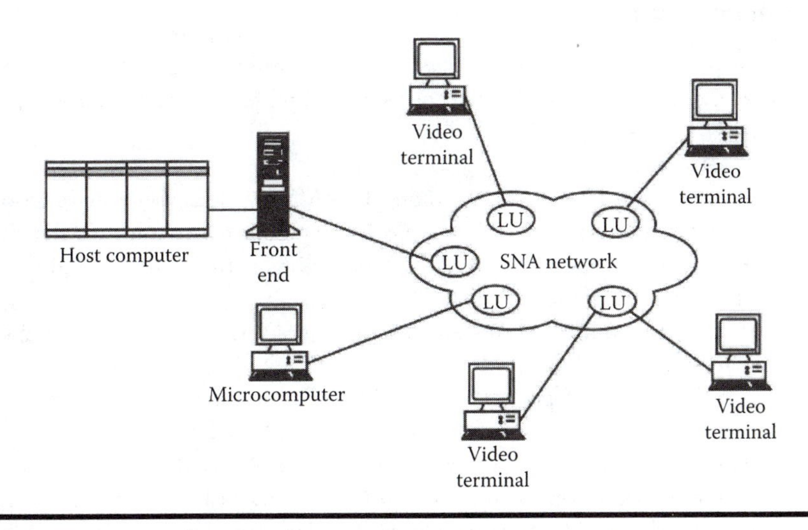

Exhibit 8.13 Synchronous Data Link Control framing session.

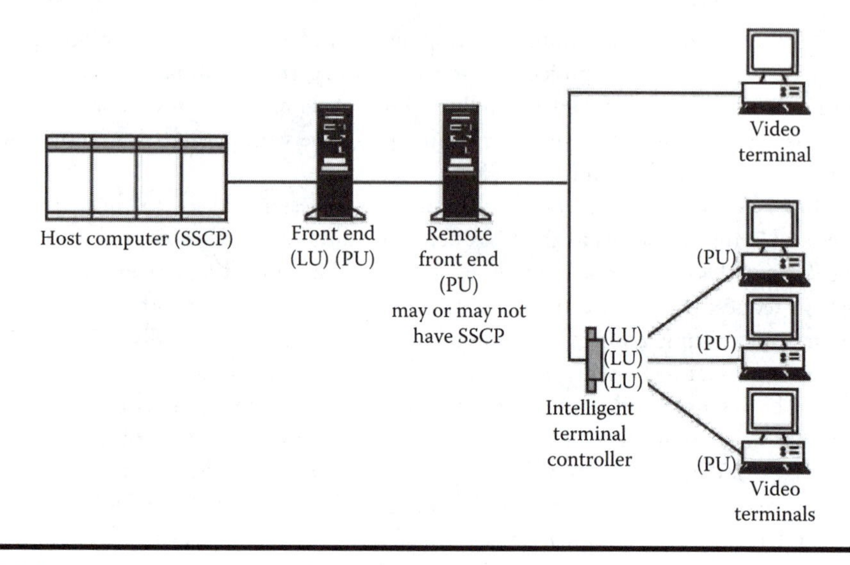

Exhibit 8.14 System network architecture device.

the routing and flow control necessary when moving data in the network. The path control mechanism handles the following:

- Prioritization of traffic
- Error detection and recovery
- Network monitoring and alerting

SNA Access Programs

Besides the previously mentioned hardware and software necessary for SNA, a telecommunications access program must be used to complete the network operation. There are four access programs available:

1. Telecommunications Access Method (TCAM): Provides the basic functions needed for controlling circuits. It has its own commands and can schedule traffic, handling operations while also providing status reporting and recovery from network failure.
2. Virtual Telecommunications Access Method (VTAM): Newer and more robust than TCAM, VTAM manages a structured SNA network.
3. Network Control Program (NCP): IBM's effort to move the TAM from the mainframe computer and distribute it to the front-end processor.
4. Customer Information Control System (CICS): A TAM that truly relieves the host computer operating system of the many tasks of telecommunications. It is the world's most widely used mainframe teleprocessing monitor,

and even the competition admits its robustness and stability. Security and message logging are also handled very well.

IBM is changing to meet the demands of the current generation. A newer SNA architecture and advanced peer-to-peer networks (APPNs) have been released. Both support industry-standard network and common data-link protocols, overcoming many of the incompatibility issues in traditional SNA.

IPv6

We have discussed that TCP/IP is one of the routed protocols of networks. The IP that we discussed was version 4, the most commonly used version of the protocol; however, there are other versions. Although IPv4 is the most commonly used version of IP, it is not without its flaws. During the growth of the Internet in the early 1990s, there were many issues that IPv4 could not handle well.

- Address space: Although designed to have adequate address space for each device, when the worldwide nature of the Internet and the commercial vendors coming on board the Internet were factored in, available addresses were rapidly being depleted.
- Security: We all want e-mail and other information transfers to be secure so that they cannot be intercepted. IPv4 does not have built-in security and is subject to being easily intercepted and stolen.
- QoS: Some messages are more important and timely than others. If we hope to use the Internet for voice and video transmission, a method must be provided to ensure that these packets are transmitted with a higher priority than e-mail (which could be minutes late without any major disruption). QoS allows for different packets to be "tagged" with codes that indicate their priority for moving through the network.

These major (and many other minor) issues led to the formation of a committee to develop IP for the next generation (IPng). Requirements were drawn up for IPng, including solutions for address space, security, and QoS issues mentioned previously. However, additional requirements were laid down. IPng must have the following characteristics:

- Be compatible and not require more robust and complicated hardware
- Have a smooth transition plan to move from IPv4 to IPv6
- Be as easy and as fast to route as IPv4

These are tall orders given that the protocol must do much more with the same resources. Many proposals were submitted, and eventually, IPv6 was accepted and ratified in the mid-1990s. Why is IPv6 not being used fully now? A fair question; the following are just two of the answers:

- Scope of change: Every PC, networking device, and end-user device must be changed from IPv4 to IPv6. This scope of change is tremendous, causing corporations to shy away from it.
- Quick fixes: Much was done to and around IPv4 that allowed it to overcome most of the major issues identified previously. For example, the built-in security of IPv6 was stolen, named IPsec, and included as an add-on to IPv4.

Eventually, the switch will have to be made from IPv4 to a newer and more robust version. Will it be IPv6? It is hard to say. What may happen is that the information technology (IT) professionals will pull out IPng and start implementing it when the public cries for more (perhaps after a critical breakdown).

Regulatory and Legal Factors

Throughout this chapter and others, we have referred to the 802.*x* standard (*x* being typically a 3 or a 5). In this section, we will explore exactly what this is.

Formally, the standard is the 802-1990 IEEE standard, put together by the Institute of Electrical and Electronic Engineers and ratified in May 1990 (hence the *-1990*). In text form, the 802-1990 standard is referred to as the *IEEE Standard for Local and Metropolitan Area Networks*. The standard was also recognized and approved by the American National Standards Institute (ANSI) in November 1990.

IEEE standards are developed by its technical committees. Members of the committees serve voluntarily and without pay, and are not necessarily members of IEEE. The standard presented represents a consensus of the experts on the subject, and use of the standard is totally voluntary. IEEE standards are subject to review at least every 5 years.

The 802-1990 standard is actually a family of standards for local and metropolitan area networks. The scope of the 802-1990 family and its relationship to the OSI Reference Model is shown in Exhibit 8.15. There are other members of the 802.*x* family, but only the most important ones are shown.

- The 802.2 part of the family describes the logical link control (LLC) medium that forms the communication link between the data-link layer and the network layer.
- The 802.3 part of the family describes the Carrier Sense Multiple Access with Collision Detection (CSMA/CD) access method and physical layer specifications for Ethernet.

OSI Model	*802 Family*			
Application				
Presentation				
Session				
Transport				
Network	802.2 Logical link			
Data link		802.3	802.5	802.11
Physical	802.1 Bridging	Medium access	Medium access	Medium access
Medium				

Exhibit 8.15 802.*x* family and the OSI model.

- The 802.5 part of the family describes the Token Ring access method and physical layer specifications for the wireless local area network.

Other parts of the family have specifications and standards for security (802.10), voice and data access networks (802.9), and metropolitan area network access and physical layer specifications (802.6).

The actual description of the standard is beyond the scope of this book; suffice it to say that the standard is complex and detailed, but it provides guidance to engineers for the development of products that can work together.

Summary

In this chapter, we have discussed in depth the subject of networks. Specifically, we have looked at the way messages get routed in a network from the origination point to the intended destination. In so doing, we have studied the components and products that the network needs.

We have looked at the types of messages that networks send: unicast, multicast, and broadcast messages. We have studied the addressing needed to send these messages, specifically the addressing used with TCP/IP.

TCP/IP uses three classes of licenses (class A, B, and C) for addressing and incorporates host ID and network ID portions of the address.

We explained the way messages move around a network through routing protocols. There are many different types of routing protocols, and their selection is dependent on the type of network and the equipment it uses.

We discussed the languages of the Internet, further exploring the structure of TCP/IP and other languages such as IPX/SPX and SNA. We ended the "Technical Factors" section of this chapter by exploring the future of IP through IPv6.

Finally, we looked at how standards are formed and why they are needed.

Review Questions

Multiple Choice

1. The three common classes of TCP/IP addresses are:
   a. D, E, F
   b. A, B, C
   c. X, Y, Z
   d. All of the above
2. CICS does which of the following?
   a. Relieves the host computer operating system of the many tasks of telecommunications
   b. Turns personal computers on and off
   c. Stores customer information in a database
   d. All of the above
3. A version number is which of the following?
   a. The version of the operating software
   b. The version of the IP language
   c. A model number located on a PC
   d. All of the above
4. Source address is which of the following?
   a. Address of a certain software package
   b. Address that specifies where a person can be found
   c. IP address from which the message originated
   d. All of the above
5. Which is not a valid IP address?
   a. 210.9.4.3
   b. 250.1.2.3
   c. 9.3.256.4
   d. 3.4.5.6
6. LU stands for:
   a. Limited unified address
   b. Logical unit
   c. Less unit
   d. None of the above
7. Which of the following is the meaning of CICS?
   a. Center for Information and Communication Sciences
   b. Customer Integration Center for Service
   c. Customer Information Control System
   d. None of the above

8. Which program is IBM's effort to move the Telecommunications Access Method from the mainframe computer and distribute it to the front-end processor?
   a. VTAM
   b. CICS
   c. NCP
   d. None of the above
9. Which are the common classes of TCP/IP addresses?
   a. A, D, F
   b. A, B, C
   c. C, D, E
   d. All of the above
10. Which of the following is accomplished by the path control mechanism?
    a. Prioritization of traffic
    b. Error detection and recovery
    c. Network monitoring and alerting
    d. All of the above

Short Essay Questions

1. What is the primary advantage of centralized routing, and why?
2. When is the SDLC framing session ended?
3. On what layer of the OSI model do the languages of the network operate, and what do they do?
4. What is TCP/IP?
5. How are multicast messages transmitted?

Chapter 9

Wireless Local Area Networks

In this chapter, we will explore the wonderful world of wireless communication. Wireless technology comes in many shapes and sizes. We use them daily: simple devices such as the TV remote control and garage door openers, and more complex devices such as cellular telephones, personal digital assistants, and satellite-based TV. Wireless is pervasive in our personal and professional lives. The scope of wireless is so wide that it would take a full book, if not a series of books, to explore the entire field.

To keep matters simple and within the scope of this book, we will explore only one application of the world of wireless: wireless local area networks (WLANs). Although WLANs are only one small part of the wireless world, they incorporate the same essential components and elements as all wireless systems; thus, the basic principles underlying wireless systems can be discussed. Additionally, WLANs function similarly to the local area networks (LANs) that we have discussed in Chapter 7. WLANs are a tremendous growth area and promise to be an integral part of our everyday personal and professional lives.

Business and Human Factors

Owing to the fast and ever-changing needs of the customer, businesses also must change quickly. You want to return your rental car fast before you catch a plane. The doctor wants to see your x-rays from the previous year. You want to check the price on a piece of clothing you are buying. All these are examples of how we need quick and readily available access to information. To accomplish this, technology must be flexible, adaptable, and easily changed. WLANs fit this bill well.

The following are the major situations that WLANs fit into:

- *The need for remote access to data*: In the warehouse, retail, and medical fields, personnel are all over the workplace and need to quickly access and enter information for customers. Examples are a retail clerk wanting to check the price of an item or a UPS delivery person wanting to verify a package location. WLANs provide access from the employee's location to the data and allow for entry and modification of data. Before the advent of WLANs, personnel had to find a wired terminal to complete these tasks.
- *Reaction to fast-changing situations*: A meeting is called, the attendees gather in a room, and they need access to the Internet as well as to data on each of their laptop computers. All of this can happen in minutes with WLANs. Additionally, people's locations between buildings and within buildings change quickly. Stringing new cable can be the answer, but the time frame for completion is too long. WLANs provide previously unheard-of mobility and flexibility.
- *Hazardous and rough situations*: Sometimes, the location or conditions do not permit ease of wired installation. For example, if a building contains asbestos, digging into floors and walls becomes expensive and lengthy. In a historic building in which modifications are not allowed, WLANs provide the solution. In a residence, where wiring is either unsightly or difficult to accomplish, a WLAN can be put into operation easily. WLANs are meant not to replace wired LANs but to supplement them in unique situations. For this purpose, WLANs make a perfect fit.

Technical Factors

Before we discuss WLANs, a bit of technical background knowledge is necessary. First, let us explore radio frequencies (RFs), specifically, the electromagnetic spectrum (EMS), and three key IEEE wireless protocols.

Basics of Radio Frequency

Almost all events in life occur in waves, specifically, sinusoidal waves. We speak in sinusoidal waves. When you throw a rock in the water, sinusoidal waves form (Exhibit 9.1). To measure these waves, we measure the time it takes for one wave to complete one cycle from start to finish. This measure of time is called cycles per second (or cps), or more commonly, hertz. (1 Hz is 1 cps; the unit is named after the person who discovered this natural sinusoidal waveform motion.)

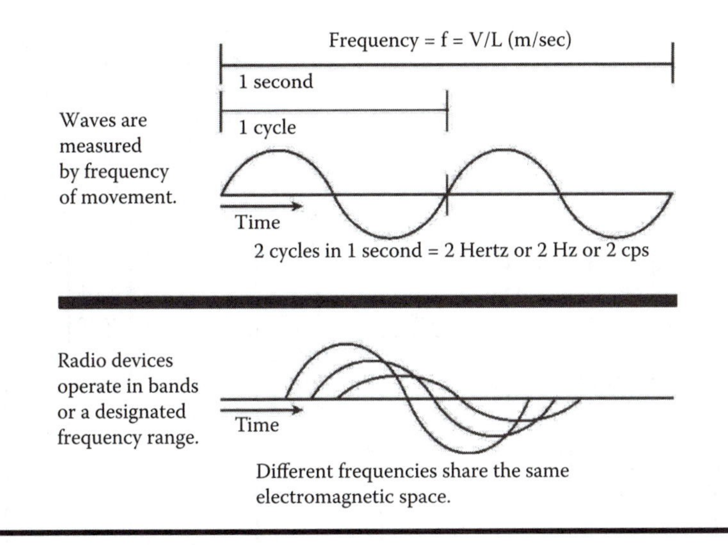

Exhibit 9.1 Sinusoidal waveform.

Sinusoidal waves behave consistently and have certain characteristics:

- The higher the frequency, the more the waves tend to behave like light. They travel in a straight line and do not bend (although they do bounce). The higher the frequency, the more the waves travel in a line-of-sight characteristic; there must be a clear path between the sending antenna and the receiving antenna.
- Electromagnetic energy in the form of sinusoidal waves decreases in power exponentially as it travels farther from the source. For example, 1 W of power 15 ft. away from the antenna is only 0.25 W of power 30 ft. away. We also see this effect in light: as we move farther away from a light source, the light becomes dimmer. This is called *attenuation*.

We can hear sinusoidal waves in the air when there is a disturbance such as a noise or someone speaks. The lower-frequency waves (around 20 cps, or 20 Hz) sound similar to low rumbles—thunder is a good example of this. As the frequency increases up to 10,000 Hz (10 kHz), we can hear human speech. Most humans can hear from about 20 to 20,000 Hz (20 kHz). Sounds of lower frequencies resemble thunder and bass drums, whereas the sounds of higher frequencies resemble whining or screeching. The waves that we can hear are at the low end of what is called the EMS.

The EMS (Exhibit 9.2) does not stop where we stop hearing. It continues to increase, and therefore, many devices can operate at (can listen to) these different frequencies. As humans, we are sensitive to another portion of the EMS: light waves. Light and sound are parts of the same spectrum, but we perceive them differently. Other devices (radios, TVs, etc.) listen and hear different parts of the EMS.

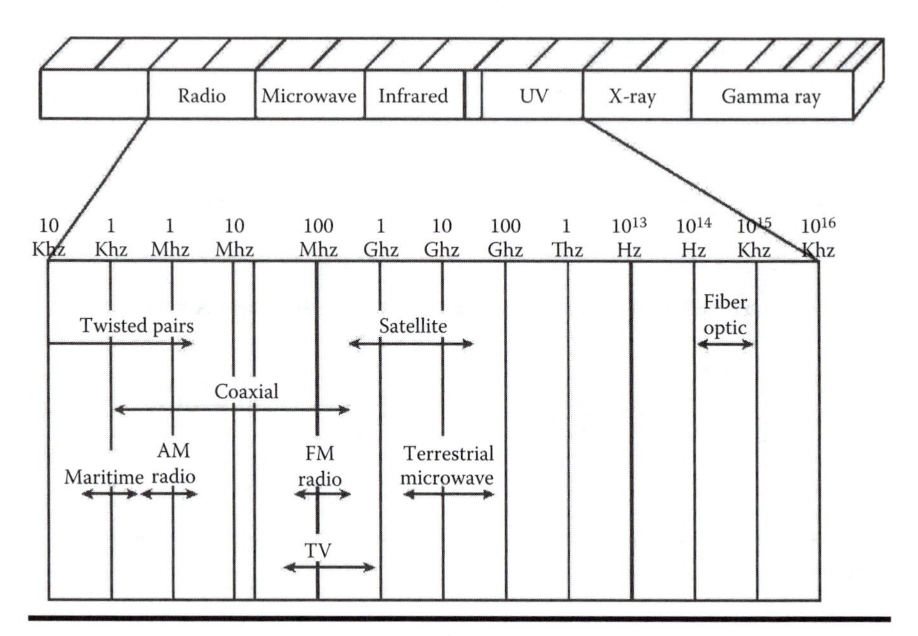

Exhibit 9.2 The electromagnetic spectrum.

Wireless Protocols

IEEE began releasing standards for wireless protocols in 1977 with the 802.11 standards family. Below are three more recent wireless standard releases.

802.11n Protocol

The 802.11n protocol for WLAN began life with discussions by the IEEE 802.11 working group in January of 2002 and was ratified by the IEEE in September of 2009. The goal of the new protocol was to increase throughput for wireless networks. The specification states that 802.11n can transmit data at 54 Mbps. The working group accomplished this with three features: multiple-input/multiple-output (MIMO) antennas, frame aggregation, and channel bonding.

Physical layer diversity (MIMO): The use of MIMO antennas has helped to increase throughput by increasing the amount of antennas used in transmitting and receiving signals. In Shannon's law, capacity increases by 1 bit every second of frequency cycle (Hz) for every doubling of radiated or output power. Capacity can be increased without doubling power by adding additional antennas to the equation. Instead of 1 bit/s/Hz, the equation becomes *N* bits/s/Hz, where *N* represents the number of antennas.

Maximum-ratio combining (MRC) is a method to select signals coming into the multiple antennas with the best signal-to-noise ratio (SNR) and amplify those signals, while attenuating signals with a low SNR. MRC is used to give more weight to stronger signals so that weaker signals do not corrupt the overall transmission with too much noise.

Frame aggregation: In earlier 802.11 standards (a/b/g), each frame that was transmitted had to be acknowledged by the receiver. While the frame was being acknowledged, no other frames could be transmitted. Frame aggregation attempts to decrease the time between sending and acknowledging of frames by combining frames that have the same physical source and destination end points and traffic class (quality of service [QoS]). The aggregated frame can be acknowledged by a single frame. In essence, multiple frames can be transmitted, and then a single frame can be sent back to acknowledge those multiple frames. This method is called Aggregated Mac Protocol Data Unit (A-MPDU). The second method, called Aggregated Mac Service Data Unit (A-MSDU), chains together frames that meet the same requirements above—the same physical end points and traffic class—but instead of combining the frames, A-MPDU chains the frames together. This creates slightly more overhead than the first method, but each frame retains its cyclical redundancy check (CRC). This means that an error in one frame only affects that single frame and does not corrupt the entire chain.

Channel bonding: Earlier 802.11 protocols used 20-MHz-wide channels for transmissions. 802.11n introduces channels that can be 40 MHz wide, which increases the amount of data that can be transmitted at one time (bit/s/Hz). Bonding channels requires the transmitting and receiving devices to use two 20 MHz channels at the same time to transfer signals. The single-antenna systems of the earlier 802.11 protocols made using wider channels impractical because of a degradation of transmission range, but the development of MIMO in 802.11n has mostly solved this problem.

Channel bonding in the 2.4 GHz range can be problematic due to other wireless protocols that operate in this space. Bluetooth, wireless handsets, and microwaves all use the 2.4 GHz band. In addition, the 2.4 GHz band only has three nonoverlapping 20 MHz channels. In the 2.4 GHz band, 40 MHz channels would overlap with each other, which would create more problems than it would solve. Bonding channels together in the 2.4 GHz band will leave fewer channels open to other devices operating in this range. The 5 GHz range, however, has the potential for 12 nonoverlapping channels. Use of 40 MHz channels in the 5 GHz range is much more prominent.

802.11ac Protocol

The IEEE 802.11ac working group began designing and developing a new WLAN standard in November of 2008. Their goal was to apply the research done into very-high-throughput (VHT) technologies to new wireless protocols. The new protocol will be able to deliver 500 Mbps to a single device and 1 Gbps total throughput for the access point. 802.11ac took the improvements of the 802.11n protocol and adapted them further for more effective use in the 5 GHz band. They accomplished this with the use of multiple-user MIMO (MU MIMO), increased modulation constellation size, wider channels, and channel bonding.

MU MIMO: 802.11n uses a single-user MIMO, which means that multiple streams can be transmitted simultaneously, but only to a single physical address. With MU MIMO, the multiple streams can still be transmitted simultaneously while supporting more than one physical address. The comparison of the two protocols can be defined as an 802.11n access point responding to network access by users as a layer 1 device or hub. It is one large collision domain, and only one user can transmit at a time. An 802.11ac access point performs similarly to a switch in the analogy. Each user path becomes its own collision domain as the access point keeps track of multiple users' transmissions simultaneously.

The MU MIMO system in 802.11ac uses a technology called null steering, which directs the specific transmission toward the correct user by increasing the power of that signal, while decreasing the power directed at the other users. For null steering to work properly, the access point needs to be constantly aware of the channels each user is on, which creates more overhead. This overhead is needed, though, because certain functions of the 802.11ac protocol, like 256-quadrature amplitude modulation (QAM), become infeasible without MU MIMO.

The MIMO functionality of 802.11ac has been expanded to support eight antennas working together. Eight transmitting antennas send streams to a receiver with eight receiving antennas. 802.11n introduced MIMO for WLAN but was restricted to four *transmit* and *receive* antennas. 802.11ac doubles the amount of antennas, which increases the amount of possible throughput. 802.11n introduced 4 bits/s/Hz; 802.11ac makes 8 bits/s/Hz a reality.

Modulation constellation size: The higher throughput rates of 802.11ac require the use of 256-QAM. The 802.11ac protocol has been rated at 500 Mbps for a single downstream link but can achieve speeds in excess of 1 Gbps when using 256-QAM. At this level, a very high SNR is required, as well as shorter distances between the transmitter and receiver. 256-QAM is restricted to a single room, as it will not work through walls.

Wide channels and channel bonding: 802.11ac employs the same techniques for channel bonding as 802.11n, but it expands the amount of channels that can be bonded together. Since it operates in the 5 GHz band, 20 MHz channels can be bonded together to form 40-, 80-, or even 160-Mhz-wide channels. Exhibit 9.3

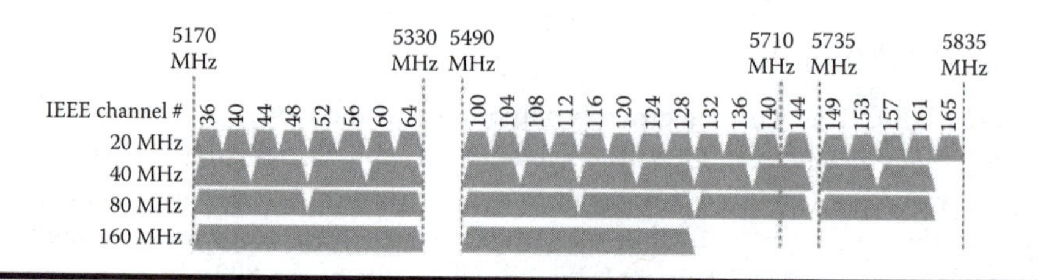

Exhibit 9.3 Channel bonding.

shows what those bonded channels would look like and which part of the 5 GHz spectrum they would use.

802.11ad Protocol

The 802.11ad protocol was developed and promoted by the Wireless Gigabit (WiGig) Alliance starting in May of 2009. The WiGig Alliance has since been absorbed into the Wi-Fi Alliance.

At 60 GHz, more spectrum is available, which means the 802.11ad protocol can take advantage of wider channels. The protocol will be able to transmit at 7 Gbps, while using low-power modulation schemes. 802.11ac can achieve similar speeds, but it requires 8 × 8 MIMO, 256-QAM, and at least four bonded 40 MHz channels. 802.11ad only needs one spatial stream, 64-QAM, and a single channel.

The new protocol will be able to support a range of powered devices, from low-power handheld devices such as cell phones to larger devices such as computers. There will also be support for other Wi-Fi protocols in the lower bands. This would allow a device to always be connected to the network while it roams from room to room. One predicated implementation of 802.11ad is replacement of the HDMI, DisplayPort, USB, and Peripheral Component Interconnect express (PCIe) wired connections. Its high throughput and numerous channels are just two features that make 802.11ad a potential contender.

What Is a WLAN?

WLANs are similar to wired LANs, the main difference being the physical medium that connects the devices. Rather than using category 5, coaxial, or fiber-optic wiring, the EMS is used. Because of this medium change, WLAN elements such as the network interface card (NIC) and the network equipment must be different. The major elements of a WLAN are (1) the access point and (2) the NIC (Exhibit 9.4).

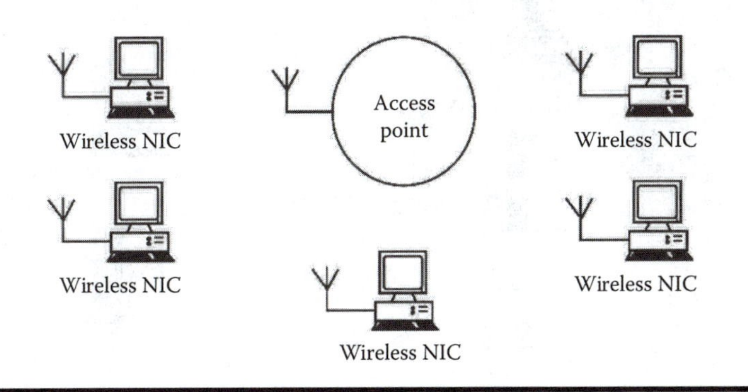

Exhibit 9.4 Elements of a WLAN.

The access point is where the wired LAN meets the WLAN. Connected to this access point is a wired connection, typically a category 5 untwisted pair (UTP) cable that allows access to all the facilities and resources available on the wired LAN. The access point has an antenna that allows entrance to the RF waves and to the WLAN. Think of an access point as a hub where each computer is connected to the network via radio waves.

The second major element of a WLAN is the NIC. Typically, the NICs are used for WLANs in laptop computers. The NIC has an antenna that allows connection to the access point. WLAN NICs are made for desktop computers also and come in varieties to match the various bus structures of desktop PC architectures. The WLAN NIC also must be matched to the operating system. For example, most WLAN NICs support Windows 2000, Windows 98, and Windows 95, and many support Linux, UNIX, and OS/2. The communication between the NIC and the operating system is enabled by drivers that the NIC vendor supplies.

A single access point can handle many computers connected to it simultaneously. Think of an access point as a hub. Typically, an access point can handle 20 to 30 WLAN NICs and their different communication sessions. A separate NIC must be installed for each computer (whether portable or desktop) to be connected to a WLAN (Exhibit 9.5).

Many WLAN vendors allow ad hoc or peer-to-peer connections. Ad hoc connection has become a standard through the IEEE 802.11 standard. In this scenario, an access point is not necessary; the NICs communicate directly with one another. This is very useful for spontaneous meetings where the participants bring their laptops with them to exchange files. Of course, resources such as the Internet are not

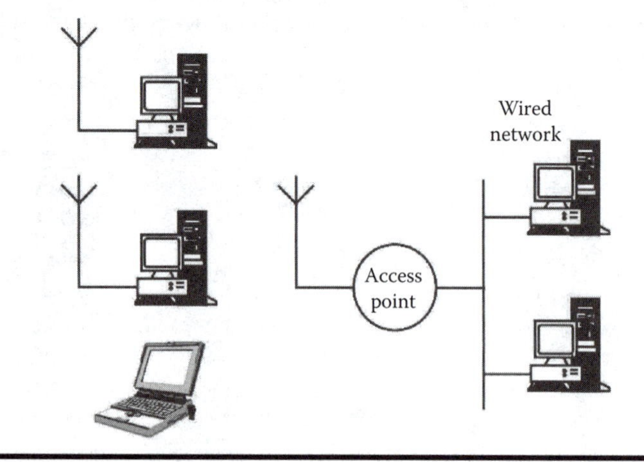

Exhibit 9.5 Wireless local area network with multiple devices.

available in this situation (no connection to the wired LAN); however, for a very low cost and no prearrangement, communication can happen.

Aspects to Consider

There are five parameters to consider when implementing a WLAN that affect its performance, operating characteristics, and, of course, the cost:

1. Access technology
2. Frequency range
3. Antenna
4. Range and throughput
5. Interference

Access Technology

Access technology refers to how the WLAN puts information onto the RF. It is a crucially important parameter as it does affect WLAN performance and its operational characteristics. Although many access technologies are used with wireless, two are predominant: (1) direct-sequence spread spectrum (DSSS) and (2) frequency-hopping spread spectrum (FHSS).

With DSSS, the packets are assigned to and sent over a specific channel (Exhibit 9.6). The channels, similar to TV channels, operate at a specific frequency; many channels can be on the same airways at any one point in time. These multiple channels allow different computers to operate at the same time without interfering with one another, similar to many people in a room all talking at once. Each person can speak at a different pitch (frequency or channel) without his or her voice mixing with that of another person.

The main advantage of DSSS is its ease of implementation and straightforward approach. Its main disadvantage is its susceptibility to interference. Because a particular channel is at a set frequency, interference at that frequency can cause the communication to be disrupted or lost. The 802.11 standard selected DSSS as its preferred access technology.

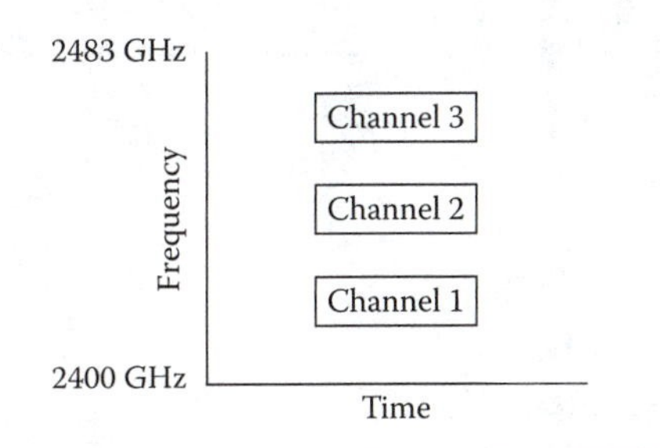

Exhibit 9.6 Direct-sequence spread spectrum operation.

With FHSS, a particular communication stream is broken down into sections, and each section is put on a different channel or frequency (Exhibit 9.7). In addition, the frequency that these sections are placed on is constantly moving (hopping) around. The sections are then reassembled at the receiving end. This is similar to speaking each of your words at a different pitch (or frequency).

This may appear to be a complex operation, but it gives FHSS its greatest advantage: resistance to interference. Because a particular communication stream is on different frequencies, it takes a large amount of interference to stop a full communication pattern. In this way, only small portions of the communication are lost, and those small parts can be transmitted again without total loss of communication.

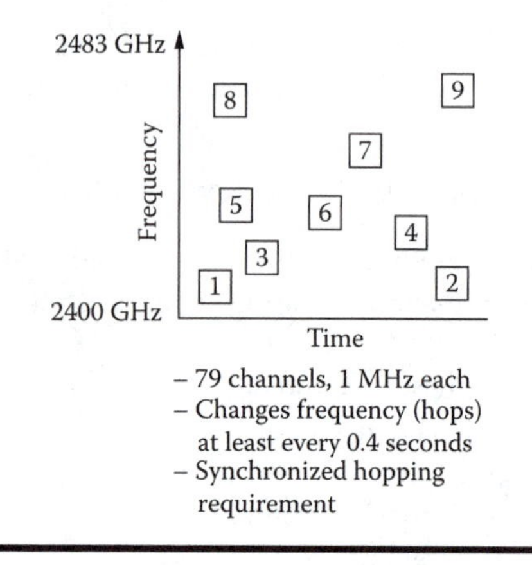

Exhibit 9.7 Frequency-hopping spread spectrum operation.

Frequency Range

The second parameter to consider is the frequency range that the product operates within. The frequency range is important because it dictates the place on the EMS where the WLAN operates. There are three common frequency ranges or bands that WLANs operate within: (1) 900 MHz, (2) 2.4 GHz, and (3) 5.5 GHz (Exhibit 9.8).

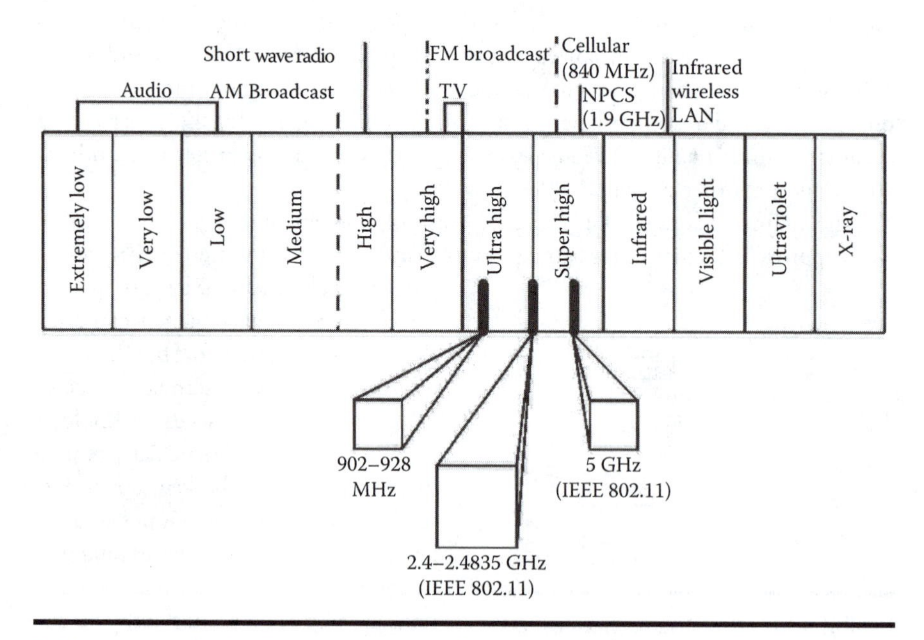

Exhibit 9.8 Wireless local area network frequency ranges.

The 900 MHz band was used first; however, today's products do not typically use this frequency. This is the lowest frequency range for WLANs, so it has the greatest operational range. The 2.4 GHz frequency range is the most popular. It provides good operational range and many different channels where both access technologies can operate.

The 5.5 GHz band is used little today; however, it will be used extensively in the future. This band provides a good degree of resistance to interference but suffers from range and line-of-sight issues.

Antennas

The third parameter, the antenna, is the physical part of the WLAN where the electrical energy starts or ends its travels through the airwaves. Antennas are used both for transmitting (sending the electrical energy to the airwaves) and receiving (taking the electrical energy from the airwaves).

There are two basic types of antennas: omnidirectional and unidirectional (Exhibit 9.9). Similar to the difference between a standard lightbulb and a floodlight, an omnidirectional antenna receives or transmits radio waves in all directions, whereas a unidirectional antenna functions only in a specific direction (measured in degrees).

Omnidirectional antennas are most prevalent with WLANs as the two antennas (the access point and the NIC) are constantly changing their relationship with each other. Omnidirectional antennas seem very useful because they transmit in a 360° spherical arrangement. However, they lose focus, called gain, because the electromagnetic energy is transmitted in all directions, including straight up, straight down, and places where almost no one will receive it (Exhibit 9.10). This is all wasted energy.

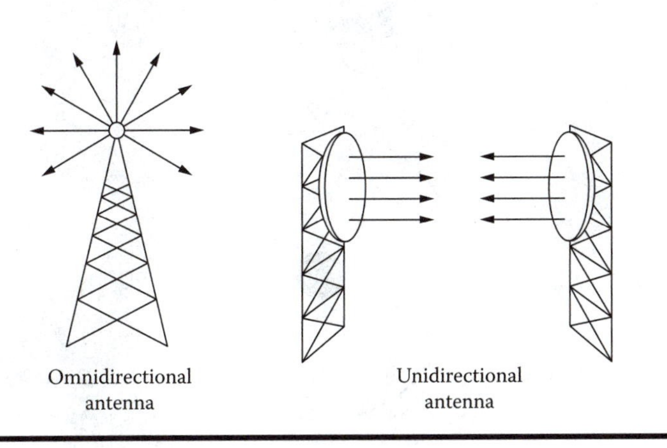

Exhibit 9.9 Omnidirectional and unidirectional antennas.

A balloon is a good example of what an omnidirectional antenna does—radiating energy in all directions equally. Squeeze the top and bottom of the balloon. What happens? The sides expand, and the balloon starts to look more like a doughnut as the air is pushed to the sides, making the sides bigger at the expense of the top and bottom. This same concept can be applied to an omnidirectional antenna, called a dipole antenna design (Exhibit 9.11). A dipole has increased gain and range on the horizontal plane. Most WLAN access points use a dipole design.

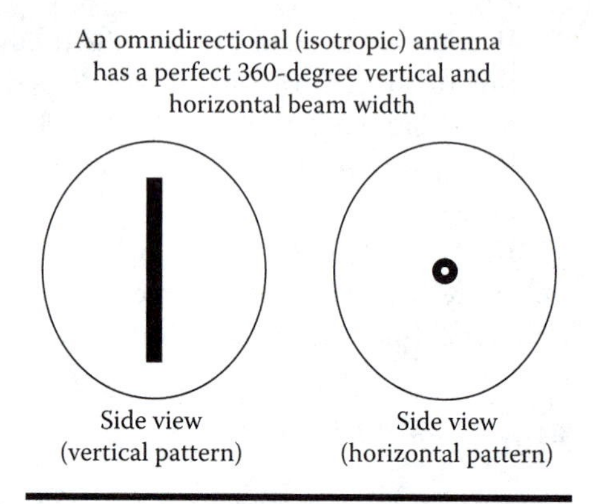

Exhibit 9.10 Omnidirectional gain.

A unidirectional antenna, similar to a floodlight, focuses the energy on the area where it is most needed, increasing its gain (Exhibit 9.12) and extending the range of the WLAN. Unidirectional antennas can be used effectively in a controlled environment.

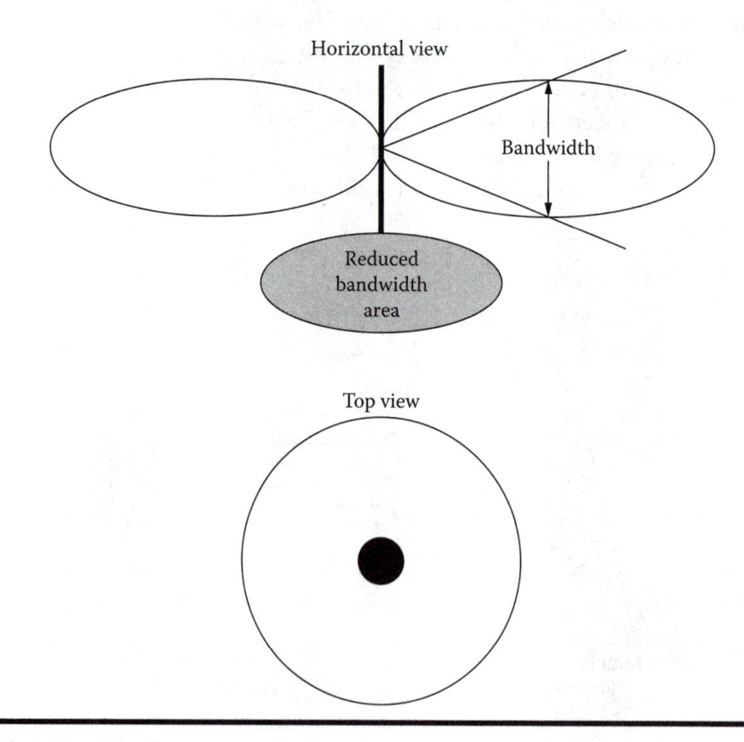

Exhibit 9.11 Dipole antenna design.

For example, students in a classroom sit in rows. If we assume that they will have their computers at their seats, we can put a unidirectional antenna in the front of the room focused on the seats. This provides a higher gain, better communication, and greater range.

For directional antennas, the lobes are pushed in a certain direction, causing the energy to be condensed in a particular area.

Very little energy is in the back side of a directional antenna.

Side view (vertical pattern)

Top view (horizontal pattern)

Exhibit 9.12 Unidirectional antenna.

Range and Bandwidth

The fourth parameter in the use of WLANs is range and bandwidth (throughput), which are grouped together because they are related to each other as well as the other parameters that we have discussed.

Range is the distance in feet or meters that the signal will travel. Similar to light, as range increases, power decreases. Throughput refers to the amount of information (packets) that can be put through (Exhibit 9.13). For comparison purposes, vendors list the maximum range of a product at the standard throughput. If a product range is given at 200 ft., you will get a full 11 Mbps throughput up to 200 ft.; beyond 200 ft., the WLAN will continue to operate, but you will get reduced bandwidth (reduced throughput). Be aware that these performance characteristics are seen under the most favorable conditions, for example, in an open field where there are no obstacles to interfere with the transmission. In a typical office environment, there are walls, windows, desks, concrete pillars, and other objects that interfere with and reduce the range of the WLAN.

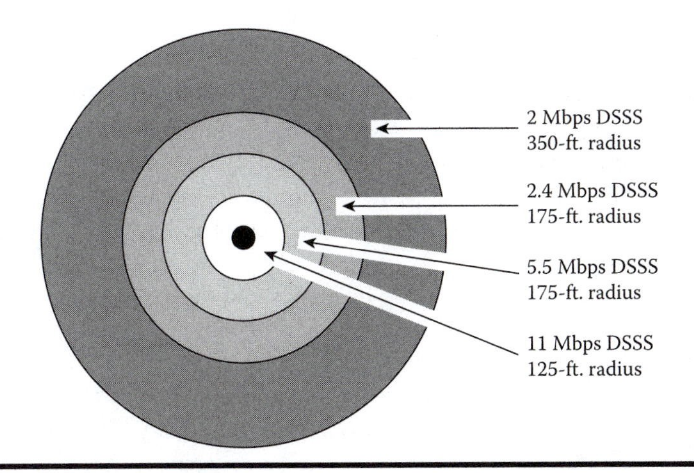

Exhibit 9.13 Range versus throughput.

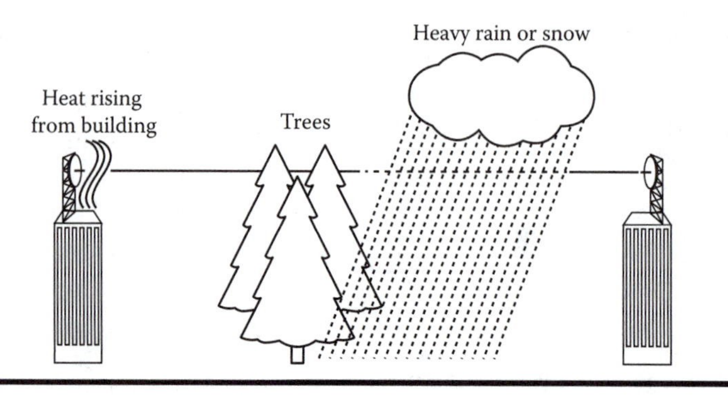

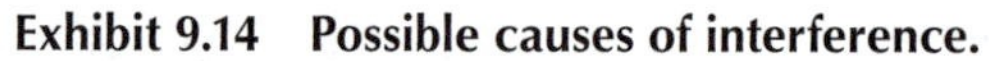

Exhibit 9.14 Possible causes of interference.

Interference

Interference refers to objects or conditions that reduce or stop the transmission of signals (Exhibit 9.14). The higher the RF, the more the waves behave as light does. RF waves travel in straight lines and bounce off objects at right angles. The frequency range of WLANs is fixed, and the two common frequencies (2.4 and 5.5 GHz) behave very much as light does. This means that all objects between the access point and the NIC cause some interference. Some objects (concrete walls, metal partitions, etc.) almost stop the transmission; other objects (wood walls, desks, or humans) just slow it down.

Interference is also caused by the operation of large motors and other power devices. Microwave ovens, heater fans, and electric motors create a band of interference that can travel quite far and interfere with WLAN transmission. Often, it is necessary to move an access point away from these types of devices to ensure proper operation.

Security

As we move from the wired LAN to the WLAN, network professionals have been forced to think more about network security to ensure that the data stay secure. They want to ensure that only authorized users are allowed on the network and that the data (now flowing around the airwaves) are not stolen or corrupted. The 802.11 standard defines a method of providing access control and privacy on WLANs via wired equivalent privacy (WEP).

WEP uses a symmetric scheme in which a key and algorithm are used to encrypt the data during transmission. The goal is to allow only authorized users on the WLAN and to ensure that the data cannot be stolen or corrupted as they travel the airwaves.

Although WEP sounds secure, most information technology (IT) professionals warn that it should not be the only means of wireless security. WEP uses a 128-bit key for encryption, and this has been shown to be vulnerable. The IEEE task group is currently working on providing more security to WEP for incorporation with future releases of the standard. The proposed new version will include different privacy algorithms and provisions for enhanced authentication.

Wireless Metropolitan Area Networks

Wireless metropolitan area networks (WMANs) connect thousands of computers together in a citywide area network. A WMAN lets you take a wireless mobile device and connect it to the Internet from any place in the geographic region where a user is located. It is a fixed wireless installation that relies on backhaul connections to the network through landline services such as traditional T-1 or T-3 circuits, optical carrier connections (e.g., OC-3), or high-throughput point-to-point wireless connections.

There are two basic types of WMANs: backhaul and last mile. Backhaul is a cost-effective solution that can be implemented in places where using fiber optics to carry user data is either cost prohibitive or, as in the case of rural markets, nonexistent. WMANs can be used in metropolitan areas and campus environments to provide the backhaul for networked services. The primary motivator for using this application in these cases is cost savings.

Last-mile solutions have the potential to establish wireless as a substitute or replacement for residential broadband digital subscriber line (DSL)/cable modems and commercial high-speed data services. Quick installation, lower cost (installation and maintenance), and the use of mature technology (radio signals have been used to communicate since Marconi!) have made WMANs a solution to provide services to markets that have been defined as cost prohibitive in deploying landline connections.

The Worldwide Interoperability for Microwave Access (WiMAX), based on the IEEE 802.16 standard, is emerging as a protocol for extending services to secondary and tertiary (rural) areas with cost parameters that will allow for extensive deployments. Current government initiatives, similar to the electrification and the extension of the telephone network to rural areas, are helping to fund a number of WiMAX rural installations. WiMAX is operated with frequencies that are licensed, unlike current Wi-Fi services. The licensed frequencies allow for greater control of user access and provide a QoS for advanced services such as Voice over Internet Protocol (VoIP) or streamed video over an IP environment.

Regulatory and Legal Factors

RF is a scarce commodity. There are only a limited number of frequencies, so they must be controlled. Each country has its own government agency for controlling RF; in the United States, it is the Federal Communications Commission (FCC). Because RFs do not stop at an artificial national border, there must also be an international coordinating agency. The International Telecommunications Union (ITU) handles international disputes or issues concerning RF.

In the United States, the FCC allocates licenses for frequencies to either the government (military, police, etc.) or commercial firms (radio stations and mobile phone companies). The FCC ensures that these transmissions do not interfere with one another. To use these controlled frequencies, the operator must obtain a license from the FCC.

The FCC has also allocated certain frequencies for industrial, scientific, and medical (ISM) use. Within these frequency bands, vendors can create devices that operate without licenses. Garage door openers and TV remote controls are examples of devices that are unlicensed because they are used within these ISM RF bands. WLAN operation falls within these ISM unlicensed frequencies. As previously mentioned, the 900 MHz, 2.4 GHz, and 5.5 GHz bands use unlicensed ISM frequency ranges.

However, even standards do not ensure that devices will operate together, because they can be interpreted differently by different people. An independent organization, the Wireless Ethernet Compatibility Association (WECA), was created to test devices to ensure their interoperability. When a product has been successfully tested, the Wi-Fi logo appears on the product box, much as the Good Housekeeping seal of approval appears on consumer goods.

Summary

In this chapter, we explored some of the business needs for wireless devices, specifically, WLANs. The ability to move data quickly and effectively, to adapt to the ever-changing environment of the business world, and to avoid hazardous spaces are all good reasons to implement WLANs.

We explored the basic elements of wireless, sinusoidal waves, and characteristics of wave motion. We looked into the basic elements of a WLAN and its operational characteristics. We discussed the NIC and access port, as well as the five operational parameters of WLANs: access technology, frequency range, antenna type, range and bandwidth, and interference.

We also looked at the agencies that control the EMS in the United States and the standards on which current WLANs are based.

Although WLANs are only a small part of the wireless world, they are typical of fundamental elements of all wireless devices. You are encouraged to explore further the wonderful world of wireless, as it will certainly be a large part of the future.

Review Questions

True/False

1. 802.3 is the body of IEEE standards for wireless connectivity.
   a. True
   b. False

2. In ad hoc mode, computers have no access to the Internet.
   a. True
   b. False
3. DSSS is more susceptible to interference than FHSS.
   a. True
   b. False
4. The access point is the physical part of any RF device where electrical energy starts or ends its travels through the airwaves.
   a. True
   b. False
5. Omnidirectional antennas have increased gain and range on the horizontal plane.
   a. True
   b. False
6. Interference refers to objects or conditions that reduce or stop the transmission of signals.
   a. True
   b. False
7. WEP has to do with ways to avoid interference.
   a. True
   b. False
8. The 900 MHz, 2.4 GHz, and 5.5 GHz bands lie in unlicensed ISM frequency ranges.
   a. True
   b. False
9. The access point jack is where the wired LAN meets the WLAN.
   a. True
   b. False
10. Because people hardly ever move, companies do not make WLAN NICs for desktop computers.
    a. True
    b. False

Multiple Choice

1. Which of the following is a reason that a company might opt for a wireless (rather than a wired) local area network?
   a. Mobility
   b. Faster setup time
   c. Easier setup procedures
   d. Building restrictions
   e. All of the above

2. One cycle per second is commonly called one what?
   a. Megahertz
   b. Furlong
   c. Megabyte
   d. Hertz
   e. Byte
3. What does NIC stand for?
   a. Network and Internet carrier
   b. National Internet center
   c. Network interface card
   d. National integration carrier
   e. Network and Internet card
4. What is the name of the hardware device that allows the wired network to connect to the wireless network?
   a. Access point
   b. Antenna port
   c. External point
   d. Eternal port
   e. Serial port
5. What is another name for a peer-to-peer connection?
   a. Ad junk
   b. Ad lib
   c. Ad nauseum
   d. Ad hoc
   e. Ad dend
6. What does DSSS stand for?
   a. Different-spectrum send simultaneously
   b. Direct-sequence spread spectrum
   c. Different-sequence spread simultaneously
   d. Direct-spectrum send sequence
   e. Different-sequence send simultaneously
7. What is the difference between DSSS and FHSS?
   a. In DSSS, the sending frequency stays the same; it changes in FHSS
   b. In DSSS, the sending amplitude stays the same; it changes in FHSS
   c. In DSSS, the sending carrier stays the same; it changes in FHSS
   d. In DSSS, the sending sequence stays the same; it changes in FHSS
   e. None of the above
8. Which of these frequency bands is not commonly used in WLANs?
   a. 900 MHz
   b. 2.4 GHz
   c. 3.2 GHz
   d. 5.5 GHz
   e. All of the above

9. Which antenna type is not commonly used in WLANs?
   a. Undirectional
   b. Omnidirectional
   c. Dipole
   d. Omnipole
   e. All of the above
10. Which IEEE standard applies to WLANs?
    a. 802.2
    b. 802.3
    c. 802.11
    d. 802.13
    e. 802.22

Short Essay Questions

1. Briefly explain a situation in which a company might choose a wireless rather than a wired LAN.
2. Explain the difference between computers in ad hoc mode and computers in infrastructure mode.
3. Describe a situation in which it might be more advantageous to have an omnidirectional rather than a unidirectional antenna.
4. Give several examples of things in an office building that would cause interference.
5. Describe some of the security issues associated with WLANs.

Chapter 10

Mobile Wireless Technologies

Wireless technologies are rapidly becoming the main delivery tool of information throughout the world. The penetration of cellular services in Europe is at 75% of the population and is moving toward total market penetration. In many European countries, a student without a cell phone attached to his or her ear or not sending text messages to a friend across the classroom is atypical. Though voice service is the major source of revenue for the cellular service providers, data applications are quickly becoming the driving force for technological change in the industry.

Three generations of wireless technologies have evolved over the past 20 years. The industry is in constant flux as it is either moving to or coming from a network upgrade or software change. Understanding the basic fundamentals of wireless technology and services will allow the reader to adapt with the industry as the next generation evolves.

Advanced Mobile Phone Service

Advanced Mobile Phone Service (AMPS) was created and tested in the early 1980s. The first test area in the United States was in the Chicago metropolitan area. From its inception, this technology has been referred to as cellular service because of the configuration of the antenna propagation field.

Although theoretically shaped like honeycomb cells on the engineering layout, the cell sites provide a way to reduce power, increase access, and reuse frequencies in the limited bandwidth allotments for cellular services. Exhibit 10.1 shows two sector layouts. The large cell site on the left has a high power output and a wide area of coverage with minimum access available. In the configuration on the right with multiple sites and different frequencies being used in each cell, the access is increased, the power of each antenna is reduced, and a similar area is being covered for service.

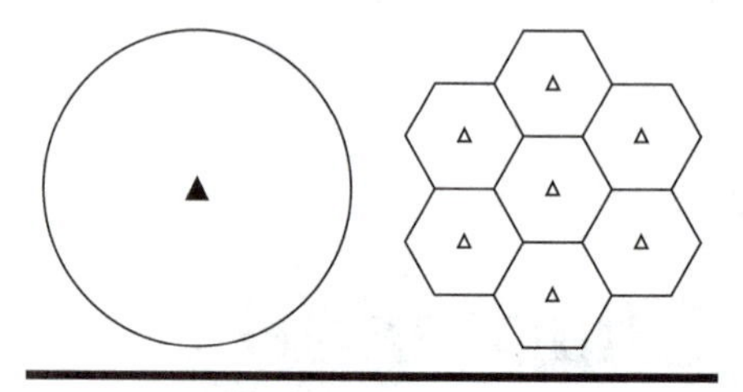

Exhibit 10.1 Cell site engineering configuration.

Because of the interrelated nature of power, antenna placement (and height), and cell radius, a radio-frequency (RF) engineer must be attentive to the number of users who may need access to the network. Early configurations of the AMPS network did not anticipate the dynamic growth of the market, which left users with busy signals in large metropolitan areas more often than their gaining access to the network.

Cutting the cells into even smaller areas, reusing the frequencies in tighter configurations, and providing for more user access addressed this issue. The reuse-of-seven rule, which states that no cell can use a frequency from an adjacent cell, forced diversity in frequency placement in the network. The bandwidth that is available to cellular providers is limited. Exhibit 10.2 shows the bands allocated to the wireless providers. Because there is a scarcity of frequency for cellular service, new and creative methods have been invented to maximize the frequencies that are available.

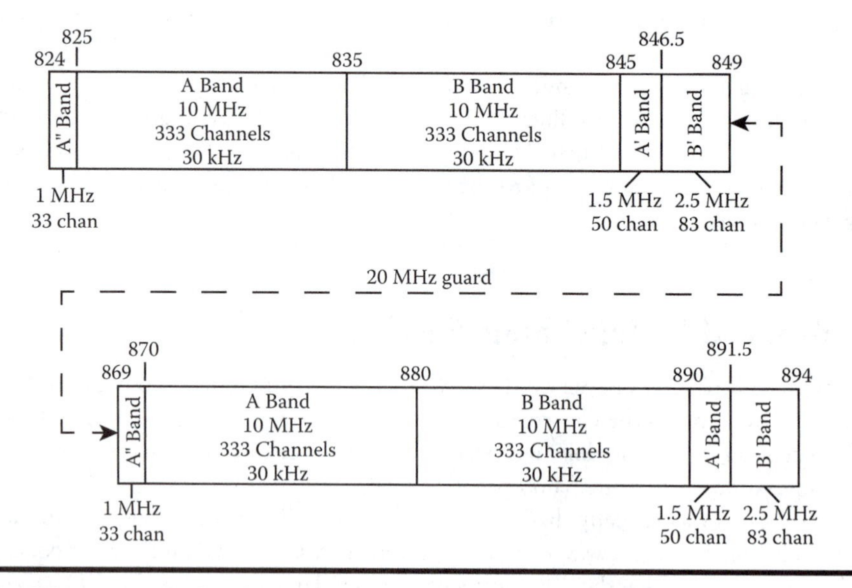

Exhibit 10.2 Wireless frequencies.

Multiple-Access Technologies

The original cellular multiple-access technology is known as frequency-division multiple access (FDMA). FDMA is a multiple-access technique in which users are allocated specific frequency bands. The user has the singular right of using the frequency band for the entire call period. Exhibit 10.3 is a graphical representation of FDMA.

Though functional and having easy-to-define frequency use, FDMA was not an efficient use of the spectrum. Because a single user captures the frequency channel until the call is completed, the maximum use of the bandwidth was not realized. FDMA has been replaced in the network with time-division multiple access (TDMA) for greater frequency use. FDMA continues to be a viable technology used in microwave and satellite transmissions, which will be covered later in this chapter.

TDMA is an assigned frequency band shared between a few users; however, each user is allowed to transmit in predetermined time slots. Hence, channelization of users in the same band is achieved through separation in time. Quiet time or unused time slots in TDMA can be maximized by inserting a user into the time slot. This method is similar to time-division multiplexing used in wire-line transmissions. Exhibit 10.4 provides an example of a TDMA configuration.

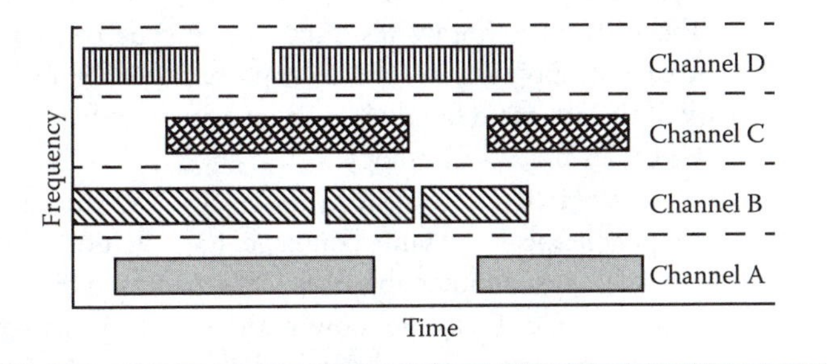

Exhibit 10.3 Frequency-division multiple access.

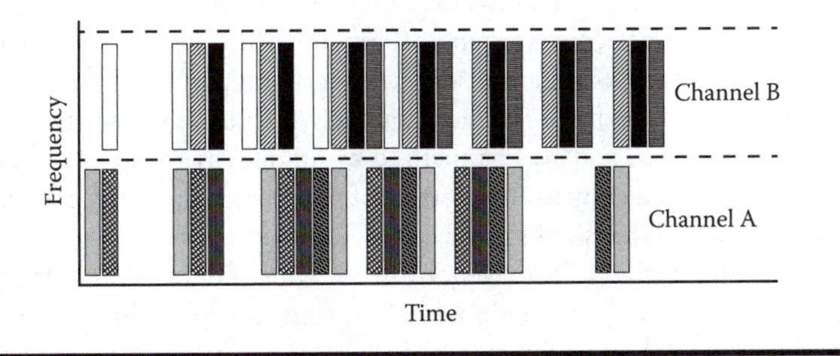

Exhibit 10.4 Time-division multiple access.

Each individual pattern in Exhibit 10.4 represents a different user at a different time in the same channel. Because there are two frequency channels allocated, there is a threefold to fivefold increase of user data over FDMA. Emerging compression techniques that compress voice bytes into fewer throughput requirements has increased the number of users with TDMA.

TDMA was developed in the United States, but it used global system for mobile communications (GSM), found in Europe and the rest of the world, as its template for development. The methodologies are similar; however, because GSM was "not invented here" (NIH syndrome), US companies were resistant to accepting the technology when networks were originally being developed. Although the United States did not accept it, GSM has the highest technology. The ability of European users to go from one country to another (roam) without disruption of cellular service is a testament to its market dominance. Because there is a singular platform for cellular services in Europe, more peripheral applications have been developed for GSM. The United States has a number of competing technologies, which makes it difficult for third-party vendors to create applications to work across all networks. This particular issue has caused the development of new wireless applications to lag behind Europe by a minimum of 24 months!

The most advanced and sophisticated method of multiple-access currently on the market was developed by Qualcomm and is called code-division multiple access (CDMA). CDMA was designed to be used by the military for secure battlefield communications. The method is highly resistant to any type of frequency jamming and electronic eavesdropping. The civilian application of the technology has boosted the capacity of the allocated bandwidth to an estimated five to seven times greater than that of TDMA-based technology.

CDMA has been analogized in the following scenario: Within a room filled with people, only two people speak the same language. Each pair of conversationalists needs to be able to talk to each other above the chatter caused by conversations in different tongues. If everyone else in the room maintains the same volume level, conversation between the two parties is possible, and all other conversation sounds like noise to them. In CDMA, the base station of the cellular network defines a specific code in which to encrypt the signal to a mobile phone. The signal is then sent out within a 1.25 MHz bandwidth along with coded information from other conversations. The other conversations, although within the same frequency band as our *define* call, are detected as noise by the receiving party and rejected. A good definition for CDMA is that it is a technique in which users engage the same time and frequency segment, and are channelized by unique assigned codes. The signals are divided at the receiver by using a correlator that accepts only signal energy from the desired channel. Undesired signals contribute only to noise.

CDMA accomplishes this delivery by using Walsh codes to provide a few million unique codes to define the signal before transmission. This level of sophistication should signal to you that a high level of signal timing, manageability, and resilience would be mandatory. CDMA also replicates the converted analog

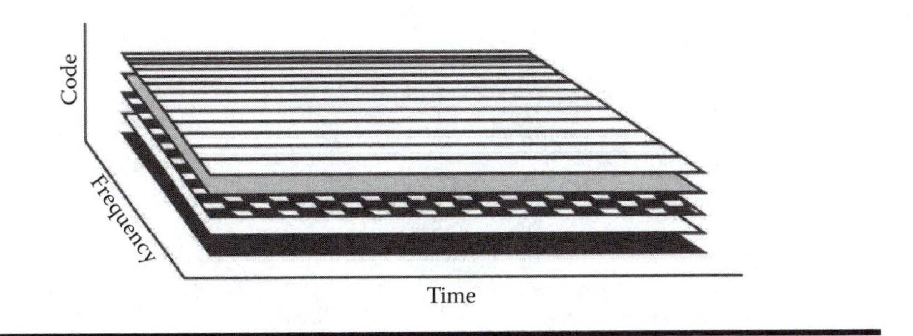

Exhibit 10.5 Code-division multiple access.

signal/digital output multiple times and spreads or interleaves the data across the entire bandwidth allocated for transmission. Along with the weaving process, the transmission randomly skips to different frequencies for brief periods of time to send the information. This bobbing and weaving process provides redundancy and security that helps deliver the information, regardless of transmission interruptions. The normal encrypted nature of CDMA technology has made it resistant to hackers who may try to steal transmissions or information relating to the identification of the handset. Exhibit 10.5 provides a visual representation of the different conversations that can occur simultaneously on the same frequency with CDMA. Each layer of Exhibit 10.5 can be visualized as a different conversation.

CDMA has the ability to deliver 10 to 20 times the capacity of FDMA for the same bandwidth. CDMA also has a capacity advantage over TDMA by five to seven times. This claim is challenged by the manufacturers and resellers of GSM and TDMA technology–based services; however, it is an evolving technology that can provide the wireless data needs for the third generation of wireless services.

Mobile Switching

The transfer of any wireless system from a mobile unit to a landline or another wireless device involves a complex path of wireless and wire-line connections. Exhibit 10.6 shows some of the major components in the network that will be discussed.

The mobile switching center (MSC), which is known by a number of different names, including mobile telephone switching office (MTSO), is the heart of the mobile network. The MSC provides the system control for the mobile base stations (MBSs) and connections back to the public switched telephone network (PSTN). The MSC is in constant communication with the MBS to coordinate handoffs (or switching) between cell sites and connections to the landlines, and to provide database information about home and visiting users on the network. The MSC acts much the same as a landline central office, except that it interfaces with the wireless component of the network also.

The MBS is the first point of contact with the mobile user. The mobile user's cellular device, while active, is constantly searching for a signal that will help it

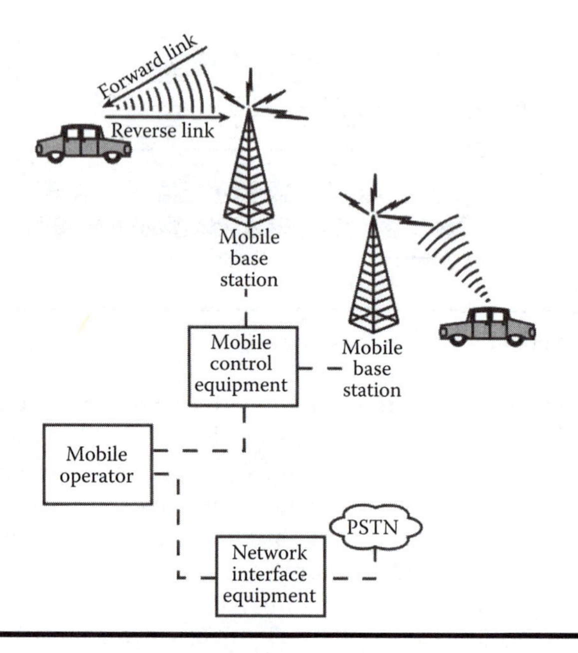

Exhibit 10.6 Mobile wireless switching.

orient to a particular MBS. The MBS is responsible for allocating available frequencies (in the case of GSM and TDMA networks) to users who travel through the cell's footprint. As the user traverses the cell, the signal strength from the MBS weakens at the perimeter of the coverage area. The user's cellular device has already been looking for another signal to switch to and has relayed that information back

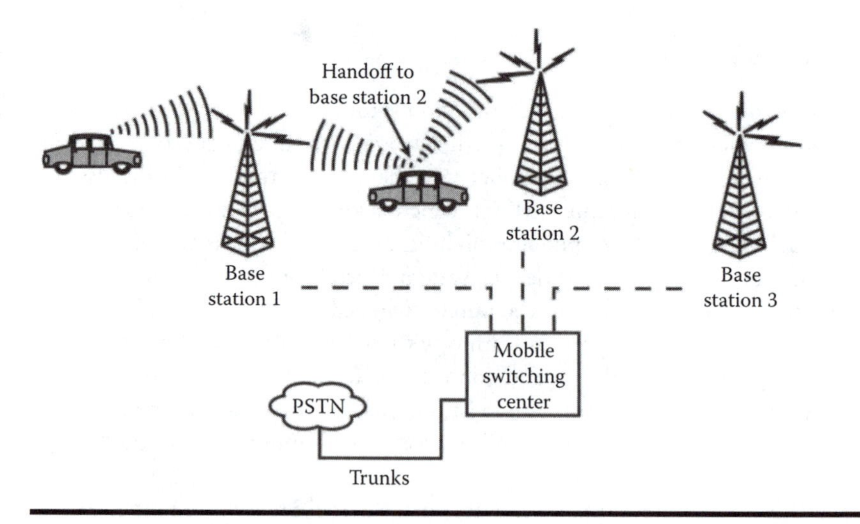

Exhibit 10.7 Mobile cellular handoff.

to the MSC through its current connection. The MSC coordinates the switch from one MBS to another with a stronger reception signal for the mobile user. Exhibit 10.7 shows how a cellular user is switched in the wireless mobile network.

Networks using CDMA technology do not look for other frequencies to pass the weakening signal to as it enters another sector. Because CDMA relies on coded signals that are broadcast on the same frequency, the handoff in this environment is directed to the stronger signal strength of the appropriate MBS. This type off handoff is termed a *soft handoff,* in contrast to that which is performed by switching frequencies in the GSM or TDMA networks' hard handoff. There are other control channels that are involved in keeping a call active (e.g., paging channel, pilot channel), but their purpose and definition are beyond the scope of this text.

Fourth-Generation Wireless (4G)

Fourth-generation wireless (4G) is the generation of wireless networks that replaced third-generation wireless (3G) networks in 2008. Standards for this technology have been set at speeds up to 100 Mbps for high-mobility communication, like trains and cars, and 1 Gbps for low-mobility communications, like pedestrians and stationary uses. 4G brings high-quality streaming video and end-to-end IP throughout mobile networks. 4G includes technologies such as software-defined radio (SDR) receivers, orthogonal frequency-division multiplexing (OFDM), orthogonal frequency-division multiple access (OFDMA), and multiple-input/multiple-output (MIMO) technologies. These delivery methods all use high rates of data transmission and packet-switched transmission protocols. Circuit-switched networks are not applied in 4G, even though they were in 3G.

Some of the forces that drove 4G wireless network deployments are price, universal access, and speed. Increased speed is a critical requirement for 4G wireless services to be successful. 4G places greater demands on the network with new wireless features, and in order to keep up with those demands, providers have increased the data rate to 10–50 times more than 3G. Doing this allows service providers to put streaming audio and video access into the hands of consumers.

Future of 4G

The future of 4G wireless technologies in the near future is bright. Countries such as the United Kingdom are just now reaching rural areas with 4G wireless technologies. The 4G wireless delivery protocol is predicted to maintain relevance until at least the year 2020, when more advanced systems will be ready for deployment. 4G will be migrated out of everyday use when the demands of users around the world (or in high-data-demand regions of the world) reach beyond the capabilities of 4G mobile networks. Although 4G may no longer be able to handle the demands of users, the infrastructure and technology will be used to migrate to fifth-generation wireless (5G), much in the same way second-generation wireless (2G, and then 2.5G) was used to move mobile service to 3G.

Future of Worldwide Interoperability for Microwave Access (WiMAX)

Worldwide Interoperability for Microwave Access (WiMAX) was the first deployed 4G wireless technology introduced to the market. WiMAX was originally designed to complement Long-Term Evolution (LTE), using both the same protocols and the same modulation schemes (e.g., OFDM). However, the rapid integration of LTE and packet switching into the mobile network by the major carriers has slowed WiMAX deployments in the major markets. AT&T and Verizon are advocates of LTE, while Sprint deployed WiMAX for its high-speed data wireless network; but recently, Sprint has decided to switch from WiMAX to LTE. Sprint switching from WiMAX has proved to be a huge hit to the technology. Mobile users are now using LTE with the three largest mobile networks in the United States. Because many technologies survive and fail by conformity to one technology over the other, many believe that the future of WiMAX is in question. To keep WiMAX relevant, many companies are combining their WiMAX networks with LTE networks to create a larger-bandwidth network. These hybrid networks will give 4G a longer life span by allowing more users to be on the network, giving the mobile environment a softer exchange into 5G wireless technologies. It should also be mentioned that WiMAX is currently not supported in the European market.

Fifth-Generation Wireless (5G)

5G is the next generation of wireless networks that will replace 4G and LTE networks in the future. Standards for this emerging technology have yet to be set. The main purpose behind 5G is to support the ever-growing number of wireless devices on mobile networks. The mobile wireless networks today carry voice and have become laden with tablets, sensors, and other devices that are causing traffic to spike on these networks. Scientists are currently working on 5G technologies that will bring 1000 times the mobile data volume than 4G wireless, setting 5G at speeds up to 1000 Gbps.

5G may include technologies such as millimeter-wave frequencies (3 to 300 GHz), smaller cell sizes, and neighbor discovery protocol (NDP). Spectrum in the 5G architecture could possibly be used as follows: 100 MHz of additional spectrum below 1 GHz to improve rural broadband, 500 MHz of additional spectrum between 1 GHz and 5 GHz to provide capacity for data, and 1.8–1.9 GHz spectrum dedicated to mobile broadband on a technology-neutral basis. It is key to remember that 5G needs to not only show improvements in speed from 4G but also be able to advance the already existing networks to handle higher traffic volumes. The issue of burgeoning data networks is the reason for adding certain amounts of spectrum to separate areas of the network.

Some of the driving forces behind 5G are the need for zero latency, the introduction of non-human-operated machines to wireless networks, and higher-capacity networks to support the current movement of the Internet of Things. 5G will need to support 10,000 times the volume traffic and 100 times the number of devices compared with today's current wireless mobile networks for future applications and demands. This new network will also need to provide 10 Gbps high data rates and 100 Mbps low data rates.

Time-Division Synchronous Code-Division Multiple Access

Time-Division Synchronous Code-Division Multiple Access (TD SCDMA) is a 3G technology standard for the physical layer that is being championed by China Academy of Telecommunications Technology (CATT). Because China is the largest cellular market in the world, the Chinese are trying to define the next generation of protocols to be used with cellular systems, not only for their own country but also on a global perspective.

The key component differentiating TD SCDMA from other forms of CDMA is the use of time-division duplexing (TDD) on the uplink and downlink from a user device. TDD shares the same frequency space rather than requiring a user to have two separate RF channels for communicating back to the base station. Synchronous time slots are allocated to provide efficient use of the frequency between the end points. When data applications are used across the TD SCDMA, asynchronous use of the time slots can be dynamically assigned to provide a greater amount of throughput for the downlink connection. Current 3G CDMA systems use frequency-division duplexing (FDD), which requires a separate channel to perform these duties. Another key difference between TD SCDMA and other forms of CDMA is that it is defined as being deployed with smart antennas. Smart antennas are next-generation devices that utilize intelligent processing to narrowly define the RF signal to each user, which allows for great frequency reuse within a cell.

Enhanced Data Rate for GSM Environment (EDGE)

Enhanced Data Rate for GSM Environment (EDGE) is a high-speed digital data service. It is a faster version of GSM, operating at a speed of 384 kbps. EDGE provides various high-bandwidth-demanding applications and broadband services on handheld devices such as multitasking mobile phones (e.g., Apple's iPhone) and computer notebooks. EDGE is constructed on the TDMA frame structure and is considered to be an evolutionary standard. EDGE technology became commercially available in 2001.

General Packet Radio Service (GPRS)

General packet radio service (GPRS) is a packet-based wireless communication service providing continuous connectivity to the Internet for mobile users. It can support data rates ranging from 56 to 114 kbps.

The channels transmit packets based on users' needs rather than allotting a dedicated channel for each user. GPRS services are therefore available to the users at a rate less expensive than what circuit-switched networks would cost. GPRS is defined for bursty data, which makes it suitable for web access, mobile email retrieval, and other applications that are not latency sensitive.

The higher data rate permits the user to take part in a videoconference or access multimedia web sites on mobile handheld devices. GPRS is widely available and supports 2.5G and 3G networks. Mobile users can activate virtual private networks (VPNs) through their GPRS networks for secure data transmissions.

GPRS is based on GSM communication and supports the existing circuit-switched mobile services and short message services (SMS). It also represents the evolution of third generation of GSM (3GSM) and wideband CDMA (WCDMA) networks.

TCP/IP runs efficiently on GPRS. It is also acceptable to x.25, a legacy landline data communication protocol in the USA and still used in Europe.

High-Speed Downlink Packet Access

High-Speed Downlink Packet Access (HSDPA) is a protocol developed for mobile telephone data transmission. The technology provides download speeds equivalent to the asynchronous digital subscriber line (ADSL) connectivity used in residential Internet access service and avoids any slow connectivity with mobile handheld devices.

The HSDPA technology is an improvement on WCDMA and 3G networks. HSDPA is associated with 3.5G networks. HSDPA works at data transfer speeds of up to 14.4 Mbps per cell for downloads and 2 Mbps per cell for uploads. However, actual applications for mobile users are more likely to experience throughput speeds of 400–700 Kbps, with bursts of up to 1 Mbps. HSDPA can still provide data rates that are at least five times faster than earlier versions of 3G data networks.

HSDPA uses different methods for modulation and coding when compared to WCDMA. It creates a new conduit or pipe within WCDMA called high-speed downlink shared channel (HS-DSCH). The HS-DSCH conduit permits faster downlink speeds. It is important to note that the connection is only used for the transmission of data to the user from the source to the mobile device. It is not possible to send data from the phone to a source using HSDPA. The connection is shared between all users, thereby allowing the radio signals to be efficiently allocated among users for fast downloads.

Since HSDPA is applicable only for high-speed data access and several other standards such as WiMAX with better potentials are gaining popularity, the long-term acceptance and survival of HSDPA have been questioned.

Antennas

The antenna is an integral part of the cellular environment. An antenna is a circuit element that provides a changeover from a signal on a transmission line to a radio wave and is used for the gathering of electromagnetic energy (i.e., incoming signals). An antenna is a passive device in the network, which means that the antenna is a receiver and transmitter of electromagnetic energy but is not responsible for amplifying the signal. Other components on the network are charged with the responsibility of making sure the signal is strong enough to be broadcast according to engineering and Federal Communications Commission (FCC) guidelines. In *transmit* systems, the RF signal is generated, amplified, modulated, and applied to the antenna. In *receive* systems, the antenna collects electromagnetic waves that are "cutting" through the antenna and brings on alternating currents that are used by the receiver. Antenna characteristics are essentially the same regardless of whether an antenna is sending or receiving electromagnetic energy.

Microwave and satellite transmissions depend on the same antenna theory to send their signals. Knowing some unique properties of antennas will help your knowledge of multiple wireless technologies other than mobile cellular services.

- *Reciprocity*: An antenna characteristic that essentially states that the antenna is the same regardless of whether it is sending or receiving electromagnetic energy.
- *Polarization*: The direction of the electric field, the same as the physical attitude of the antenna (e.g., a vertical antenna transmits a vertically polarized wave). The *receive* and *transmit* antennas must have the same polarization.
- *Radiation field*: The RF field that is created around the antenna and has specific properties that affect the signal transmission.
- *Antenna gain*: The measure in decibels of how much more power an antenna will radiate in a certain direction with respect to that which would be radiated by a reference antenna.

Antennas come in various sizes and shapes and have specific functions based on the RF signal reception characteristics. Antennas are designed primarily on the wavelength of the signal to be transmitted and received. Half-wave (1/2) and quarter-wave (1/4) antennas use a simple formula that defines their size in meters or in feet (Exhibit 10.8).

The radiated pattern of the antenna depends on a number of factors. Antennas can be designed to meet specific broadcast requirements and focus the radiated beam

$$\lambda = c/f = \frac{186{,}000 \text{ miles/second}}{\text{frequency of the signal}}$$

where
c = speed of light
λ = wavelength of the signal
λ use 3×10^8 when dealing in meters for the speed of light

Exhibit 10.8 Wavelength formula.

of RF energy in a narrow pattern rather than being sent in an omnidirectional pattern (Exhibit 10.9). Directional antennas are used to direct the RF energy toward a specific geographic region. The power of the signal drops off at a fairly rapid rate away from the centerline of the signal. The power of the transmission drops to half (–3 dB) at points on either side of the main focus of the signal. The width between these two power drop points is called the beamwidth (Exhibit 10.10).

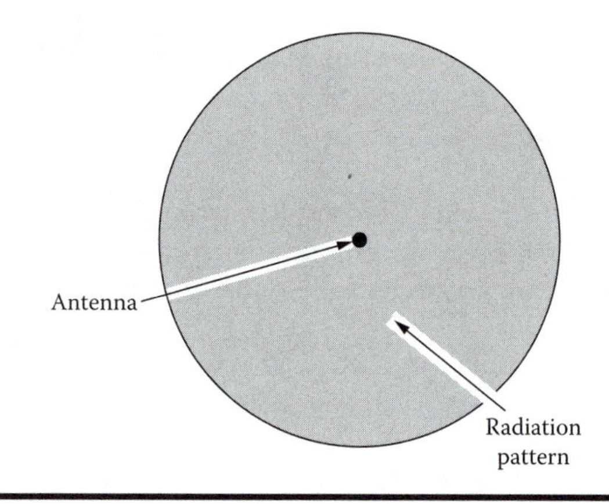

Exhibit 10.9 Omnidirectional antenna: radiated power pattern.

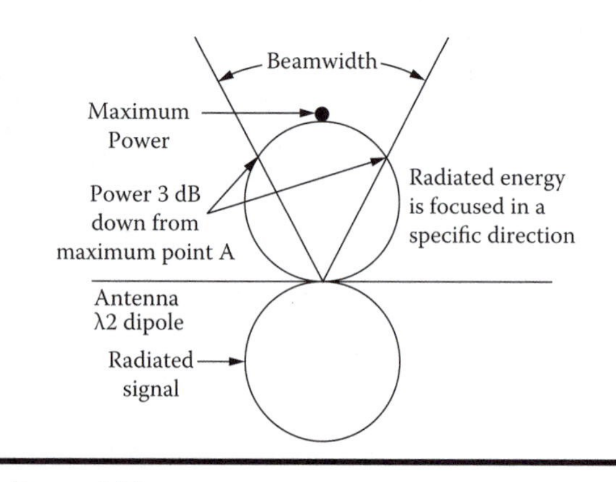

Exhibit 10.10 Beamwidth.

Smart Antennas

Smart antennas are base station antennas with a pattern that is not fixed but, rather, adapts to the current radio conditions. With a smart antenna's ability to change the direction of its radiation patterns, performance characteristics of a wireless system can increase in capacity. The goal of a smart antenna is to expand the radio-based signal quality system through more focused transmission of radio signals while maximizing capacity through increased frequency reuse.

The idea of smart antennas can be visualized as the antenna directing a beam toward the communication link only. Smart antennas add a new way of separating users, namely, by space, through *space division multiple access* (SDMA). By maximizing the antenna gain in the desired direction and simultaneously placing a minimal radiation pattern in the directions of the frequencies that hinder the signal frequency, the quality of the communication link can be significantly improved. SDMA implies that more than one user can be allocated to the same physical communications channel simultaneously in the same cell, only separated by angle. In a TDMA system, two users will be allocated to the same time slot and carrier frequency at the same time and in the same cell. In systems providing full SDMA, there will be much more within-cell or *intracell handovers* than in conventional TDMA or CDMA systems, and more monitoring by the network will be necessary.

There are two major categories of smart antennas with regard to the transmission strategies. The first type is the *switched lobe* (SL), which is also called the *switched beam*. The simplest smart antenna technique, the SL system forms multiple fixed beams that are extremely sensitive in specific directions. They detect signal strength, choose one of the fixed beams, and switch from one to another as the mobile moves throughout the sector. SL systems take the outputs of multiple antennas and combine them to form narrow-sectored beams with more selectivity than any single antenna system in regard to the space the frequency is being broadcast into.

The second category of smart antennas is the *adaptive array* (AA). Using a variety of new signal-processing algorithms, this system is able to effectively locate and track different types of signals in order to minimize the interference and maximize the intended signal reception. This adaptive system provides optimal gain while identifying, tracking, and minimizing interfering signals. These techniques will take full advantage of the *signal-to-interference ratio* (SIR) for better transmission quality.

Some of the salient features of smart antennas are as follows:

- Signal gain: for efficient coverage of a given level of signal and for optimal power usage, inputs from multiple antennas are combined.
- Interference rejection: in order to improve the signal-to-noise/interference ratio, the antennas can be directed to sources experiencing cochannel interference.

- Spatial diversity: the effects of fading and multipath propagation are minimal.
- Power efficiency: the gain in the downlink channel is optimal as the result of combining multiple elements in the input signal.

Along with advantages such as increased capacity, signal quality, and coverage, smart antennas are also advantageous on two other fronts. Smart antennas support multipath rejection, i.e., higher bit rates are possible without the use of an equalizer. Smart antennas are also cost-effective; they are low on amplifier costs and power consumption and high on reliability.

Wireless communications are rapidly becoming the most common solution for voice and data transfers due to the mobility and flexibility of wireless systems. The current wireless systems cannot compete with some of the services offered by a wired line due to limited data rates. However, there is a growing demand for better access for customers with reduced infrastructure costs. The best solution is to use smart antennas, more specifically, MIMO technology.

MIMO is a promising technology for the next generation of wireless technologies, especially in the fixed-broadband (e.g., WiMAX 802.16) environment. By using the parallel transmission of frequencies created by the multiple-antenna sending and receiving configuration, the throughput for wireless systems can be greatly increased.

Microwave Signals

The knowledge of RF signal propagation, which was discussed in the chapter on modulation (Chapter 3), and the information on antennas within this chapter are the fundamental building blocks necessary to understand how microwave signals are transmitted and received. Microwave signals continue to be used today in land-based systems and are the medium of choice when communicating with satellites. The range for microwave signals is from 1 to 40 GHz. To understand how much information can be transmitted within this huge bandwidth, consider that all the AM and FM radio, broadcast television, aviation communication, public safety (e.g., police radio), and most mobile cellular channels reside under the 1 GHz frequency! A bandwidth of 1 GHz represents a huge space in which information can be transferred, while microwave is 40 times that amount.

Microwave transmissions were developed during World War II and were converted for civilian use as long-haul carriers for frequency-division multiplexed trunks across the United States. Fiber-optic cable infrastructure has replaced most of this service because fiber has a higher quality of service (QoS), thanks to a lower bit error rate (BER), which means that the number of pieces of information corrupted or lost during transmission is better with fiber-optic versus microwave transmission. Microwave continues to be a viable means of signal transmission for right-of-way issues, geographic obstruction problems, and wireless-local-loop

(WLL) considerations. Impairments to microwave signal transmission include the following:

- Equipment, antenna, and waveguide failures
- Fading and distortion from multipath reflections
- Absorption from rain, fog, and other atmospheric conditions
- Interference from other frequencies

These issues will be discussed in more detail later in this chapter. The impairment questions need to be asked during any engineering configuration for a microwave transmission facility. These issues include the following:

- Free space and atmospheric attenuation are defined by the loss of signal traveling through the atmosphere. Changes in air density and absorption by atmospheric particles are the principal factors affecting the microwave signal in a free air space.
- Reflections cause multipath conditions and can occur as the microwave signal traverses a body of water or fog bank.
- Diffraction is the result of variations in the terrain the signal crosses.
- Rain attenuation or raindrop absorption is the scattering of the microwave signal, which can cause signal loss in transmissions.
- Skin effect is the concept that high-frequency energy travels only on the outside skin of a conductor and does not penetrate into it any great distance. It determines the properties of microwave signals.
- Line of sight (LOS) is defined by the Fresnel zone. Fresnel zone clearance is the minimum clearance over obstacles that the signal needs to be sent. Reflection or path bending occurs if the clearance is not sufficient (Exhibit 10.11).
- Fading is caused by multipath signals and heavy rains, which can cause signal disruption. Exhibit 10.12 is a depiction of how the microwave normal and faded signals are viewed. High frequencies are repeated and received at or below 1 mi. Lower frequencies can travel up to 100 mi., but 25 to 30 mi. is the typical placement for signal repeaters.
- Range is the distance a signal travels. Its increase in frequency and its extended range are inversely proportional repeaters. Back-to-back antennas and reflectors are used to extend the reach of microwave signals.
- Interference is as the name implies. There are two types of interference to be considered. The first, adjacent channel interference, is caused by signal transmissions of frequencies too close in proximity. With the advent of analog-to-digital signal conversion, digital technology is not greatly affected by adjacent channel interference. The second type of interference is overreach. Overreach is caused by a signal feeding past a repeater (or *receive* antenna)

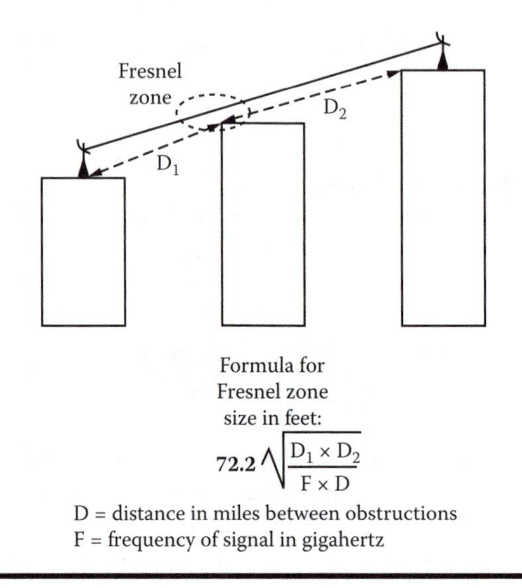

Exhibit 10.11 Fresnel zone considerations.

to the receiving antenna at the next station in the route. This condition is eliminated by zigzag path alignment or alternate frequency use between adjacent stations.

Each of these specific considerations must be assessed prior to system installation or addressed by your RF vendor to satisfy QoS considerations prior to adopting a microwave transmission solution.

The components of a microwave system are comprised of three basic elements: the digital modem, the RF unit, and the antenna. The RF unit is based on properties similar to other radio signal modulation schemes. You can start to visualize the building blocks of the previous technologies discussed, helping to prepare you to understand this technology.

The digital modem modulates the information signal (intermediate frequency [IF]) into the RF unit. From there, the RF unit pushes the signal through to the antenna. A direct connection to the antenna is preferred. The RF unit connects to the antenna with either coaxial cable or a waveguide, a hollow channel made of a low-loss material used to guide the RF signal to the antenna for broadcasting. The primary methods used today for modulating the signal for transmission are AM and FM signaling.

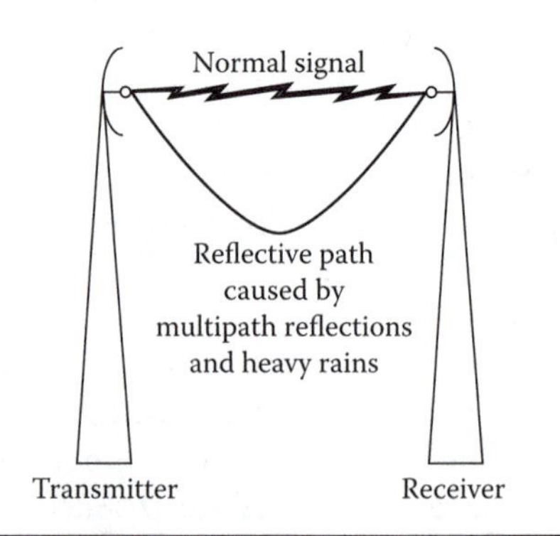

Exhibit 10.12 Fading.

Engineering Issues for Microwave Signaling

Because RFs begin to behave in a manner similar to a beam of light as the frequency increases (moves up the spectrum), there are some specific issues that must be addressed when planning microwave transmission links. The first consideration is the allowable BER for the transmission. The BER is a performance measure of microwave signaling throughput. A count of more than 10^{-6} (one error per million transmitted bits of information) is usually considered an acceptable bottom threshold for data transmissions. If the data failover is at 10^{-3} (one error per thousand transmitted bits of information), voice traffic can withstand this higher error rate.

There are four primary diversity issues that must be considered with microwave placement: space diversity, frequency diversity, hot standby, and the use of a Primary Rate Interface (PRI) connection for a failover. Space diversity protects against multipath fading by automatically switching over to another antenna placed below the primary antenna on the *receive* site. This is done at the BER failure point or signal strength attenuation point to the secondary antenna that is receiving the transmitted signal at a stronger power rating (Exhibit 10.13).

Frequency diversity uses separate frequencies (dual transmit-and-receive systems); it monitors the primary signal transmission for failover at a specific BER and switches to the standby or redundant frequency. Interference usually affects only one range of frequencies, so switching from the primary to the secondary signal usually resolves the interference problem. This type of diversity is not allowed in noncarrier applications, because of spectrum scarceness (Exhibit 10.14).

Hot standby is designed for equipment failure only. This method provides a complete redundant set of transmission and reception gear for the site. This is an expensive but necessary proposition if there is a critical nature associated with the link. Exhibit 10.15 gives a block diagram of what a hot standby microwave transmission link contains.

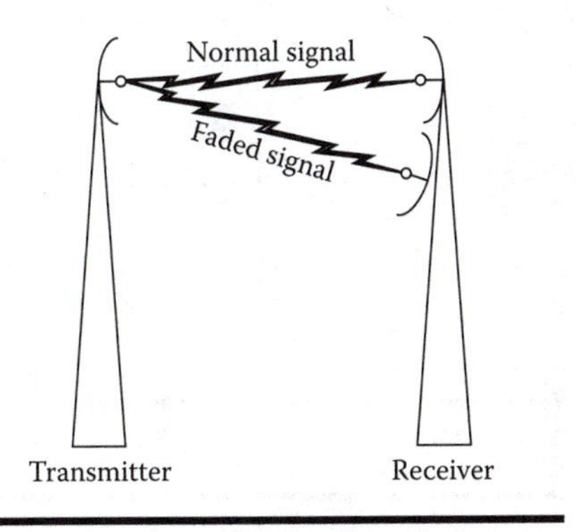

Exhibit 10.13 Space diversity.

Using an Integrated Services Digital Network (ISDN) PRI circuit as a failover for exceeding the BER for a particular signal transmission is a cost-effective method of providing a redundant path for the transmission. Though the PRI circuit cannot handle the same throughput as the microwave signal, it would provide for a minimal redundant connection, while the primary connection regains an acceptable BER, or in the case of equipment failure, until the defective parts can be replaced (Exhibit 10.16).

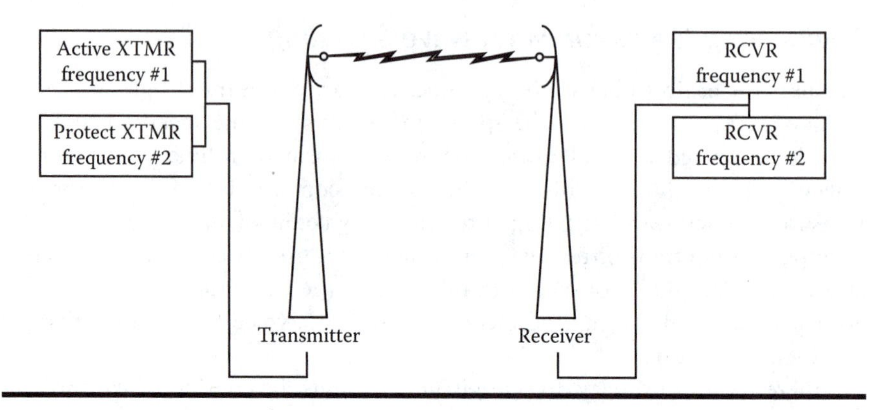

Exhibit 10.14 Frequency diversity.

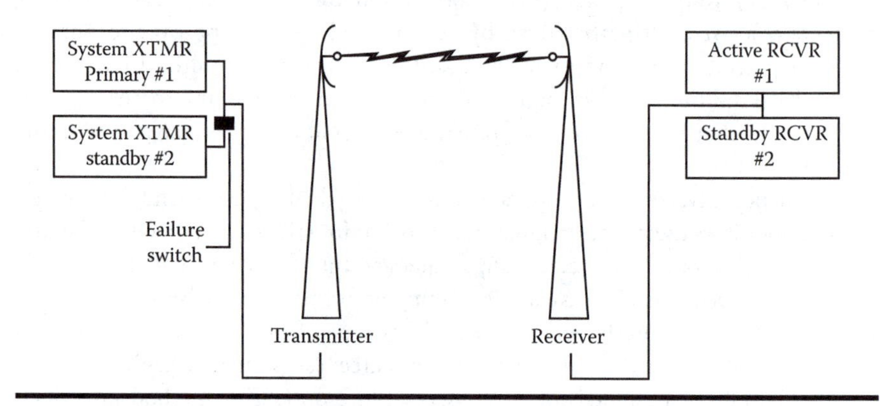

Exhibit 10.15 Hot standby.

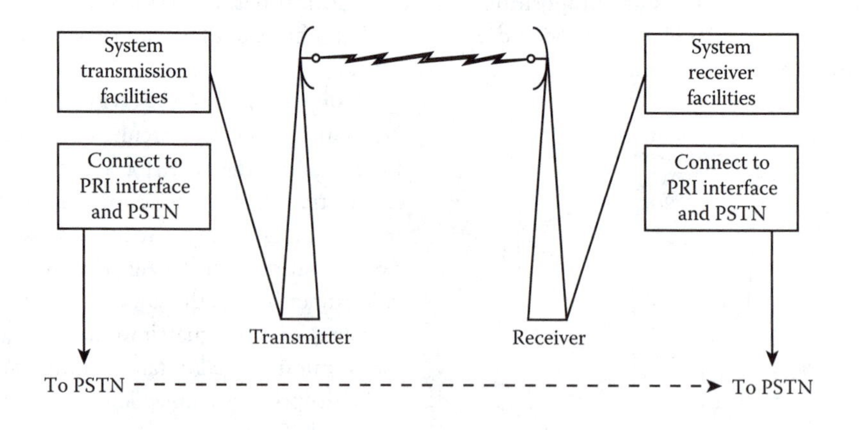

Exhibit 10.16 PRI failover.

Percent availability = 1 – (outage hours/8760 hours per year)
Private microwaves have 99.99% availability

Exhibit 10.17 Availability formula.

A final consideration that is constructive in planning a wireless broadband connection is to estimate the mean time between failures (MTBF) of the equipment being used. All electronic equipment will fail eventually. The manufacturer provides an estimated time frame in hours for the durability of the equipment. A simple formula is used that can help calculate quickly from the engineering specifications sheet how reliable the equipment will be for the particular application. Exhibit 10.17 gives the basic mathematical computation for the availability formula.

Summary

Wireless technologies are capable of providing new and ever-changing applications for the business environment. The migration from the original AMPS technology for mobile cellular service to FDMA, TDMA, GSM, and CDMA has created greater opportunities with the available bandwidth allocated to service providers.

A widely developed application of the wireless technologies is the WLL. WLL is a system that substitutes the copper cable that connects the subscriber's premises to the PSTN through radio signals. The WLL market already counts million of users. It is being developed in third-world countries as a substitute to the plain old telephone system (POTS) because it is usually cheaper to install MBSs rather than connect the entire premises with copper cable. It is also developed to promote broadband in rural areas.

WLL can be implemented on different analog and digital cellular networks. Analog cellular networks, such as AMPS, are limited in terms of capacity and functionality; however, they are expected to be a major platform in the short term because of their widespread implementation. As digital cellular networks replace analog cellular ones for mobile services, digital cellular (GSM, TDMA, CDMA, and WiMAX) is the emerging standard for WLL, especially WiMAX, which offers the highest capacity and a high level of privacy thanks to a licensed, secured spectrum. Personal communications networks also have their role to play by allowing the user to be connected to the telephone network regardless of its location.

The technology exists to implement WLL intensively, and the deployment of a new network is a matter of months; however, deploying these systems costs comparably less than wire-line solutions, and return on investment is realized sooner, especially in competitive environments.

Exhibit 10.18 Case study: wireless local loop.

Antennas are a critical part of any wireless network. Their purpose in the system is to transmit and receive the signal in various formats. Next-generation smart antennas are being developed to provide additional use of the limited frequencies between and within particular cells. SDMA is emerging as a unique form of multiple access for antenna frequency reuse.

Microwave technologies have been around since World War II. They were developed into the primary long-haul mechanism for telephone services. Although they have been replaced by fiber-optic trunking for this service today, they have developed niche services for WLL, telephone company bypass, and short-haul connectivity in metropolitan environments (Exhibit 10.18).

Review Questions

Multiple Choice

1. What is the advantage of cell site engineering?
   a. Reduce power
   b. Increase access
   c. Reuse bandwidth
   d. All of the above
2. What is an advantage of GSM over TDMA?
   a. It was invented in Europe
   b. It is technically better than TDMA
   c. It has a higher technology penetration in the world
   d. All of the above
3. What is the wavelength formula?
   a. $\lambda = c \times f$
   b. $\lambda = f/c$
   c. $\lambda = c/f$
   d. None of the above
4. What is the availability formula? Percent availability equals:
   a. 1 – (outage hours × 8760)
   b. 1 – (outage hours/8760)
   c. 1 + (outage hours/8760)
   d. None of the above
5. In which type of handoff is the user assigned a new frequency?
   a. Soft handoff
   b. Medium handoff
   c. Hard handoff
   d. All of the above

6. The power of the transmission drops in half (–3 dB) at points on either side of the main focus of the signal. What is the width between these two power drop points called?
   a. Beamwidth
   b. Radiation pattern
   c. Wavelength
   d. None of the above
7. Impairments to microwave signal transmission include:
   a. Fading and distortion from multipath reflections
   b. Absorption from rain, fog, and other atmospheric conditions
   c. Interference from other frequencies
   d. All of the above
8. What are the two types of interference for a microwave signal?
   a. Adjacent channel interference
   b. Reflection
   c. Overreach
   d. Diffraction
9. Which of these components is part of a microwave system?
   a. Digital modem
   b. Analog amplifier
   c. Laser unit
   d. Waveguide or coaxial cable
10. What is the bit error rate (BER)?
   a. The threshold under which a mobile user is disconnected from the network
   b. A performance measure of microwave signaling throughput
   c. Indicates that a hacker corrupts the transmission channel
   d. All of the above

Matching Questions

Match the following terms with the best answer.

Match the wireless frequencies:

1. A-band uplink	a. 880 to 890 MHz
2. A-band downlink	b. 825 to 835 MHz
3. B-band uplink	c. 835 to 845 MHz
4. B-band downlink	d. 870 to 880 MHz

Match the definition:

1. FDMA	a. Users engage the same time and frequency segment and are channelized by unique assigned codes.
2. TDMA	b. Users can be allocated to the same physical communications channel simultaneously in the same cell, only separated by angle.
3. CDMA	c. Assigned frequency band shared among a few users.
4. SDMA	d. Users are allocated specific frequency bands.

Match the increase of users:

1. TDMA	a. 5 to 7 times over TDMA
2. CDMA	b. Depends on the type of antenna
3. CDMA	c. 3 to 5 times over FDMA
4. SDMA	d. 10 to 20 times over FDMA

Match the definition:

1. Fourth-generation wireless (4G)	a. Wireless networks designed to support the ever-growing number of wireless devices on mobile networks at speeds up to 1000 Gbps.
2. Fifth-generation wireless (5G)	b. Wireless networks that enabled high-quality streaming audio and video at speeds up to 1 Gbps.
3. Long-Term Evolution (LTE)	c. The first deployed 4G wireless technology, not currently supported in the European market.
4. Worldwide Interoperability for Microwave Access (WiMAX)	d. The three largest mobile network providers in the United States currently provide this 4G technology.

Match the definition:

1. Reciprocity
2. Polarization
3. Radiation field
4. Antenna gain

a. The antenna is the same regardless of whether it is sending or receiving electromagnetic energy.
b. The measure in decibels of how much more power an antenna will radiate in a certain direction with respect to that which would be radiated by a reference antenna.
c. The direction of the electric field, the same as the physical attitude of the antenna.
d. The RF field that is created around the antenna and has specific properties that affect the signal transmission.

Match the antenna length by using the wavelength formula:

1. f = 1 MHz, antenna length = $\lambda/4$
2. f = 100 MHz, antenna length = $\lambda/2$
3. f = 1 GHz, antenna length = $\lambda/4$
4. f = 10 GHz, antenna length = $\lambda/2$

a. 15×10^{-3} m
b. 75×10^{-3} m
c. 1.5×10^{-3} km
d. 46.5×10^{-3} mi.

Match the definition:

1. Switched-lobe antenna
2. Phased-array antenna
3. Omnidirectional antenna
4. Adaptive-array antenna

a. Continuous tracking can be accomplished.
b. The radiation pattern can then be adjusted to void out interference.
c. Contains only a basic switching operation between separate directive antennas or predefined beams of an array.
d. Not a smart antenna.

Match the definition:

1. Atmospheric attenuation
2. Reflections
3. Diffraction
4. Raindrop absorption

a. The scattering of the microwave signal
b. Defined by the loss the signal undergoes traveling through the atmosphere
c. Can occur as the microwave signal traverses a body of water or fog bank
d. The result of variations in the terrain the signal crosses

Match the definition:

1. Skin affect
2. Line of sight
3. Fading
4. Range

a. Caused by multipath signals and heavy rains
b. Defined by the Fresnel zone
c. The distance a signal travels
d. The concept that high-frequency energy travels only on the outside skin of a conductor and does not penetrate into it any great distance

Match the definition:

1. Space diversity
2. Frequency diversity
3. Hot standby
4. PRI connection

a. Protects against multipath fading by automatically switching over to another antenna placed below the primary antenna on the *receive* site.
b. Transmission is a cost-effective method of providing a redundant path for the transmission.
c. Monitors the primary signal transmission for failover at a specific BER and switches to the standby or redundant frequency.
d. Provides a complete redundant set of transmissions and reception gear for the site.

Match the definition:

1. BER $> 10^{-3}$
2. BER $< 10^{-3}$
3. BER $< 10^{-6}$

a. Acceptable for both voice and data
b. Not acceptable for both voice and data
c. Acceptable for voice only

Short Essay Questions

1. Define AMPS and explain the reuse-of-seven rule.
2. Define FDMA.
3. Define TDMA.
4. Define CDMA.
5. Define MSC and explain the two different types of handoffs.
6. Define *antenna* and name its four unique properties.
7. Define *smart antenna* and name three types of smart antennas.
8. Define SDMA.
9. Define microwave signals and give their range of frequencies.
10. Define BER, the four types of diversities, and MTBF.

Chapter 11

Virtualization

Compute virtualization is a technique of separating the physical hardware of a computer from the operating system (OS). This then enables multiple OSs to run at the same time on a single machine. Moreover, this process can further be extended to a cluster or pool of machines running many OSs and can give the impression to each OS that it has the complete physical machine or cluster of machines to itself. Each OS can then manage and allocate shared resources, without knowledge that such sharing extends way beyond the set of applications that each OS serves. Virtualization thus encapsulates an OS and a set of applications into a logical entity termed a *virtual machine* (VM).

Demand for virtualization has risen as the workforce increasingly connect their laptops, tablets, and smartphones to applications residing in virtual data centers (VDCs), freeing the users to access office databases from home. This chapter will first discuss the process of virtualizing the computer and its utilization. We will then discuss virtualizing data storage and data networking infrastructure. Finally, we will examine virtualization of the workstation and applications in the mobile environment.

Compute Virtualization

A VM is not a *real* or physical computer. It is a logical processor that appears to be and behaves like it is a real, physical machine. Placing an OS on a VM enables a set of applications to run as if they were on their own physical machine. More than one OS (with its own supported applications) can run on a single VM. Furthermore, many of these VMs can then be placed on one physical computer, significantly increasing the usage capacity of a single physical machine.

In compute virtualization, a software-based virtualization layer resides between the hardware and the VM on which an OS is running. The virtualization layer is implemented by a software package called the hypervisor.

Hypervisor

A hypervisor software package allows many OSs, each with its own applications, to run on a single physical machine. The purpose of the hypervisor software is to provide all the VMs access to the hardware resources, including the central processing unit (CPU), internal memory, and an internal network. The hypervisor interacts directly with the physical resources of a computer system and serves as an intermediary between the hosted OSs and the applications that run over them. This delays the cost of hardware acquisition by more efficiently making use of physical computers through the sharing of computing resources, and is a key component in effective data center consolidation efforts.

Hypervisors are generally used in one of two ways, as a bare-metal hypervisor or as a hosted hypervisor. A bare-metal hypervisor is directly installed on the computer hardware and has direct access to all the hardware's resources. Exhibit 11.1 illustrates the role of a bare-metal hypervisor in a virtualized computer. A hosted hypervisor is installed and run as an application on top of an OS. Since the hosted hypervisor is running on an OS, it supports the broadest range of possible hardware configurations that the OS can handle.

The two key components of a hypervisor are the kernel module and the virtual machine monitor (VMM).

Kernel Module

The kernel module provides the core OS to hardware functionalities, such as hardware resource scheduling and input/output (I/O) stacks for reading and writing to external devices such as a keyboard, mouse, CDs, thumb drives, etc. The kernel module establishes a path for communication between the applications on VM's

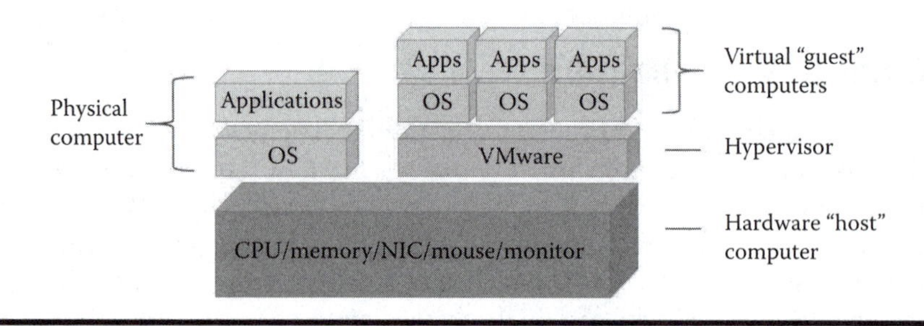

Exhibit 11.1 Bare-metal hypervisor supporting three virtual machines.

OS and the hardware and I/O devices. The hypervisor's kernel module also provides *process* creation and scheduling, as well as file system management. A *process* is the official term for a loaded program, including its code, identity information, preliminary page table, queue entries, and the stack information used by the OS.

Virtual Machine Monitor

The hypervisor's VMM actually executes each hosted application's commands on the CPU. Each VM is assigned to the VMM, which is allocated a portion of the CPU, memory, and I/O devices on the physical computer. The VMM allows the shared hardware to appear to the OSs and their applications as an unshared physical machine with its own CPU, memory, and I/O devices (keyboard, mouse, external storage devices, etc.). When a VM starts running, the control is transferred to the VMM, which subsequently begins executing instructions from both the application and the hosting OS assigned to that VM processing through the hypervisor on the physical computer. The VMM also performs binary translation (BT) for software such as Java-encoded programs, which might contain some instructions that are not directly supported by the hosting computer's hardware.

The process of compute virtualization is very effective in optimizing the standard hardware resources of a physical machine. However, optimization is limited by the extent of that hardware's computing power. One means of expanding the resources available to a physical machine is virtualized storage.

Virtualized Storage

Storage virtualization is the process of hosting storage arrangements by presenting one or more logical storage units to the accessing computer. These logical storage units are really subcomponents of physical storage resources the computer is connected to over the data center fiber network. But to the VM applications accessing it, the virtual storage appears to be located directly in the physical machine. The hypervisor can access either directly attached storage or network-attached storage (NAS). The latter may be over a Fibre Channel (FC) network to an FC storage device or over an Internet Protocol (IP) network to IP storage devices such as internet small computer system interface (iSCSI) or NAS devices. VMs running through the hypervisor remain unaware of the total storage space available to the hypervisor and the underlying storage technologies that are employed. They only see the virtual aspects of that portion of the storage that has been identified to be *virtually available* to them. The logical to physical storage mapping is performed by storage virtualization layer software running on the virtual computer. This virtual storage layer abstracts the identity of physical storage devices and creates a *storage pool* of aggregated storage resources from multiple storage arrays across the network. Storage can be added to or removed from the storage pool without affecting application availability. This provides flexibility in dynamically allocating components of the real storage and effectively spreads the storage load to minimize the purchase of an increasing amount of storage units.

Virtual Machine–Based Storage Options

Virtualization applied at the network-wide level enables the pooling of storage purchased from many vendors and the ability to manage these pools from a single management interface. It also allows data to be moved back and forth between these arrays of storage quickly. When virtualizing storage over a network, the network-based storage virtualization process places the virtualized network storage resources on the hypervisor-hosted computer. It then provides an abstract view to the hypervisor of the physical storage resources. So when an I/O command is issued from an application program running on a VM, the command is sent through the OS, to the hypervisor, and on to the processing computer, and is then redirected through the network's virtualization layer to the mapped physical storage.

There are two types of network virtualization approaches: block-level storage and file-level storage. Block-level virtualization is used for a storage area network (SAN), an environment of storage devices external to the local area network (LAN). File-level virtualization is used for NAS, storage devices internal to a LAN.

Block-Level Storage

Virtualization at the block level is an abstraction layer placed between the physical storage resources in a SAN. Instead of connecting individual servers to specific groups of physical machines, the storage of multiple servers is collected into a logical storage pool that is virtually connected to many VMs. This approach to network-based storage virtualization enables dynamic resource allocation of a heterogeneous set of storage devices and arrays.

When you have employed block-level virtualization, the virtualization appliance in the network also performs any required migration of data. This software allows storage volumes to remain accessible by users even while data migration is taking place. One reason for such data migration is to provide for lower-valued data to be migrated from high-performance storage to lower-cost, lower-performance storage in order to free up preferred storage equipment.

File-Level Storage

File-level storage virtualization, in contrast to block-level virtualization, occurs in an NAS environment. Storage virtualization at the file level provides abstraction in the process of separating files from their actual physical locations on physical disks. Before file-level virtualization, each client knows the exact location of its file-level resources. But in a data center, the migration of data from one NAS device to another is sometimes needed to refresh technology, meet performance requirements, and add storage capacity. File-level virtualization simplifies file mobility by creating a logical pool of storage, enabling users to employ a logical path in place of a physical path to access files. This approach facilitates the movement of files between the NAS systems without the normally required downtime, and clients can access their files even while the files are being migrated.

In order to map the logical path of a file to the physical path, *global namespaces* are used. Namespaces provide the abstraction layer, which enables clients to use logical names that are independent of actual physical locations. With a standard file system, a namespace is associated with either a single machine or a file system. A global namespace provides a single view of all directories and files by bringing multiple file systems under a single, common namespace and providing a single control point for managing files for administrators.

Virtual Provisioning and Automated Storage Tiering

Storage administrations need to plan in advance for the usage of the storage under their management control. Storage space must be purchased, installed, and allocated in advance based upon the cumulative amount of each client's forecast of their growth and future processing needs. This process is a difficult balancing act. Overpurchasing results in earlier and sometimes higher costs, increases the required amount of power and cooling, and causes the occupation of more precious floor space. Furthermore, even when the appropriate storage is purchased, poor planning can lead to misallocation to the wrong, less needful clients. This balancing of needs, costs, and just-in-time availability is the core activity of storage administration.

One solution to storage balance is virtual provisioning of thin pools and thin LUNs. Deployment of thin pools and thin logical unit numbers (LUNs) is best suited for situations where it is difficult to predict the impending storage space consumption of certain applications. The anticipation is that more efficient utilization of storage will occur by reducing the amount of static storage assigned to individual computer systems that often waits to be used and then may be used only briefly.

Thin Pools and Thin Logical Units (LUNs)

A thin pool is a collection of physical storage drives in a SAN or NAS environment. Thin LUNs are logical drives of storage created out of a thin pool and allocated a computer system. All the thin LUNs created from a thin pool share the same storage resources of that pool. From the thin LUN's shared pool of storage resources, required amounts of storage capacity are then allocated to each application on demand as needed. When thin LUNs are no longer needed by a computer system, they can be logically removed, and the physical storage capacity can be returned back to the thin pool. Rebalancing is performed after the pool formally reclaims the allocated capacity. The rebalancing reallocates the storage on physical disk drives over the entire pool, ensuring that the capacity of each disk drive is uniform across the pool. Multiple thin pools can be created within a storage array. Virtual provisioning simplifies the provisioning of storage to computer systems by making it only virtually available, independent of the physical storage capacity that has been provisioned. The time required to repeatedly add storage capacity to the computer systems can be reduced, and storage can be separately provisioned in a virtual fashion. The overall physical capacity can be planned and then provisioned based on the current needs of all the clients in use. This improves capacity utilization by reducing the amount of allocated but unused physical storage and also avoids overallocation of storage to the computer systems. As a result, virtual provisioning reduces storage and operating costs of the storage provider.

Virtualizing the Workstation

With the traditional desktop, the OS, applications, and user data are all tied to a specific piece of hardware. With legacy desktops, business productivity is impacted greatly when an end-point device is broken or lost, because the user has lost the use of the OS, the applications, the user data, and all the settings. Even if the hardware is replaced, without the data, the user's functionality is still lost.

In a virtualized desktop, virtualization breaks the bonds between hardware and these critical elements, enabling the information technology (IT) staff to change, update, and deploy the various hardware elements independently for greater business agility and improved response time. End users also benefit from virtualization because they get the same desktop but with the added ability to access the computing environment from different kinds of devices and access points in the office, at home, or on the road.

Computer-based virtualization separates the OS from the hardware but does not inherently address the broader and more important aspects of separating the user's data, settings, and critical applications from the OSs themselves. This is very critical for maintaining a productive business environment. By separating the user's data, settings, and critical applications from OSs, organizations can begin to achieve a true IT–business aligned strategy. This is the fundamental concept underlying virtualized desktops. Application virtualization, discussed in a later section, further breaks the dependency between the application and the underlying platform that includes the OS and hardware.

The next step in centralizing computer system resources is desktop and application virtualization. These techniques gather the dispersed resources of numerous machines and devices into one centrally managed environment.

Benefits of Desktop Virtualization

Manageability: One of the major concerns of an organization is management of the desktop infrastructure. This is because the core component of a desktop infrastructure is the personal computer (PC) hardware, which continues to evolve rapidly.

Organizations are forced to deal with a variety of hardware models, face complicated PC refresh cycles, and handle support problems related to PC hardware incompatibilities with applications. This requires that IT personnel spend most of their time supporting the existing environments instead of deploying new technology. Virtual desktops are far easier to maintain than traditional PCs. Because of the unique characteristics of VMs, they provide simple flexibility to the user when accessing. Since the application physically resides in the data center, it is much simpler to patch/fix/repair applications, provision new users, remove users, and migrate the applications to new OSs, new computers, and even new data centers.

Security: Applications and data stored in traditional desktops pose severe security threats, especially if the desktops/laptops are lost or stolen. Damage due to data loss can be immeasurable. Compliance regulations (e.g., Sarbanes–Oxley, Health Insurance Portability and Accountability Act [HIPAA]) protect privacy and corporate data by restricting the public disclosure of these data when issues arise. Failure to protect these data can lead to a significant negative impact on the organization's reputation. Since virtualized desktops are accessed by smartphones and the like and run as VMs within a remote and protected data center, this mitigates the risk of data leakage and theft and simplifies compliance procedures. Since the centralized virtual desktops reside entirely within the data center, it is easier to ensure full compliance with backup and recovery policies.

Cost: The cost of deploying and managing a traditional PC is high and ranges up to $350, approaching the cost of the device itself. The objective of desktop virtualization technology is to centralize the PC OS at the data center. This is to make security and management of desktops significantly easier. Desktops hosted at the data center run as VMs within the VDC, while end users remotely access these desktops from a variety of end-point devices. Application execution and data storage do not happen at the end-point devices; everything is done centrally at the data center.

Mobility: A key benefit of desktop virtualization in this highly mobile world is the ability to use thin clients as end-point devices. This creates an opportunity to significantly reduce the cost of user personal hardware by replacing desktop and laptop PCs with thin-client devices such as smartphones and tablets, which tend to have a life span twice that of a standard PC. Furthermore, these thin clients consume less power when compared to standard PCs, and they can be carried and used almost anywhere. This is especially useful in situations where users need to work from home, on trains and planes, in customer offices, and away from their desks in a variety of other remote worker scenarios.

Desktop Virtualization Techniques

There are two main methods of desktop virtualization from a VDC environment: remote desktop services (RDSs) and virtual desktop infrastructure (VDI). For the end user, the two are indistinguishable. The distinction is in the provisioning of the VDC resources.

Remote Desktop Service

RDS is a mature process that has been widely deployed through Microsoft. End users install an RDS client on their personal device that connects remotely to a shared VM in a VDC. All RDS users connect to the same VM OS. Individual application sessions are distributed from the shared VM's OS to multiple users' devices. These sessions mimic a desktop experience in that the clients remotely receive the visual feedback of the session on their mobile devices, while resource consumption actually occurs on a server in a pool of such servers in a remote VDC.

Virtual Desktop Infrastructure

VDI is the most widely used desktop virtualization technology for managing customer desktop applications from a virtualized data center and will be explored in detail here.

VDI involves the hosting of a desktop OS running on a VM, which itself operates on a virtual sever in a VDC. The user's device is connected to the virtual desktop by a connection broker. The connection broker is responsible for establishing and managing the network connection between the end-point device and the desktop VM that is running on the VM-hosting server. Once a remote connection is established, the user has full access to the resources of the virtualized desktop even though it is hosted and operates remotely in a VDC and is accessed by means of a network.

VDI Benefits

Improved deployment and management: Virtualizing the desktop and decoupling the OS environment from the physical hardware of the desktop, laptop, or mobile device enables rapid desktop provisioning and deployment of new applications, additional users, and services. Since the desktop OS environment is centralized on managed hosting servers in the data center, the hosted desktop environments can be more effectively managed. Software updates, fixes, and upgrades can be completed centrally instead of on each user's end-point device.

Improved security: VDI also provides security benefits for the desktop environment by storing and running the desktop OS environment on centralized servers located in the data center, instead of on distributed desktops or end-point devices where there is more opportunity for unwanted access and for the devices to be hacked, stolen, or lost. If the desktop or laptop is lost or stolen, the user's desktop environment is still protected. Centralization of the desktop environment on servers in the data center, employment of good management practices, and employing well-constructed security software that is integrated with the service delivery software have the strong potential to make for stronger security in general; they can centrally update and test the security system and the operational plan, and make implementation and testing of security software updates and the overall security of the data center operation more efficient.

Better business continuity and disaster recovery: Implementing VDI can improve desktop business continuity and disaster recovery since it centralizes the desktop OS, the user's applications, and the user's data in the managed data center, where administration can more easily be performed and backup and recovery operations are formalized, documented, practiced, and employed on a regular basis by trained employees.

VDI Limitations and Considerations

Network connectivity: The limitations of VDI are that it relies on networks to connect to the data center–based applications and that a network connection must be available and transmit at the prescribed level of performance. If the user's end-point device cannot connect to the hosting server, the user will be unable to access his or her desktop, applications, and data.

Additional infrastructure required: A VDI solution requires additional servers, storage, and networking infrastructure depending on the scale, performance, and service-level requirements of an organization. This may require new servers and added storage into the data center if an organization is planning to support a large number of dedicated users with their VDI solution.

Virtualizing the Application

Virtualization of an organization's application software is becoming increasingly popular. This approach aggregates OS resources within a virtualized container along with the application that accesses and uses them. Thus, the organization can deploy applications without modifying or making any change to the underlying OS, file system, or registry of the computing platform in which they are deployed. Since virtualized applications run in an isolated VM environment, the underlying OS and other applications are protected from potential corruptions and disruptions due to installation and modifications. Conflicts often arise if multiple versions of an application are installed on the same computing platform. In some cases, two versions of a Microsoft product cannot be installed on the same computing platform at the same time. However, when you virtualize one or both of them, they can now both be used simultaneously. The two methods that are commonly used for deploying application virtualization are application encapsulation and application streaming.

Application Encapsulation

Application encapsulation packages the application in a self-contained executable package that does not rely on a software installation or an underlying OS for any dependencies. This package is accessible from the network on a USB key or via local storage. Because these applications have the capability to function like standalone executables, they do not require any agent to be installed locally in the client machine where they run. (Built-in agents are present within the package.)

Application Streaming

Application streaming is the process of sending to the user's device only a minimum (20–30%) of the total application data when an application request is submitted and the application is about to be executed. Thus, the start-up of the application occurs quite quickly, and the effect on the network is reduced. Additional features of the application are delivered when demanded by the user and then operate in the background without direct user intervention. All application software is centrally stored in the VDC on the pooled storage array. When sufficient network bandwidth is available to the user's device, application streaming is a well-performing application approach.

In conclusion, application virtualization simplifies the deployment of new applications and the modification of existing ones for the using customer, who avoids the process of installing each application on a particular machine with the appropriate OS, and simplifies the process of changing those applications and OSs. Furthermore, due to the separation of the application from a specific OS, the management of OS images is simplified, and OS patches and upgrades are hidden from the customer and the customer's applications. This creates an environment in which the desktop appears to the user as a collection of separately managed components, each with its own virtual OS resources. This simplifies the situation for the client user while eliminating resource contention and application conflicts as well.

Summary

Traditional desktops have significant management and maintenance overhead and also lack security if they are lost or stolen. Virtualization of storage devices, workstations, and applications is an effective solution to these problems.

In this chapter, we discussed the basic structure of compute virtualization and the role of the hypervisor. Next we covered the two methods of storage virtualization, block-level and file-level virtualization, and the use of thin pools and LUNs in storage allocation. The two desktop virtualization techniques are RDS and VDI. In RDS, a service runs on top of a single Windows installation and provides individual sessions to client systems. VDI refers to hosting a desktop OS running in a VM on a server at the VDC. VDI is widely used when compared to RDS in a VDC environment. Application virtualization enables running an application without installation or any dependencies on the underlying computing platform. Application streaming and application encapsulation are the commonly used methods for deploying application virtualization. Thus, virtualization of the computer, the storage, the accessing network, and the user's desktop provides consistency, efficiency, scalability, manageability, and security in operating these endowments separately as individual physical entities.

Review Questions

True/False

1. Virtualization is a technique of separating the physical hardware from the operating system.
   a. True
   b. False
2. The purpose of the hypervisor is to provide standardized software resources.
   a. True
   b. False
3. The two key components of the hypervisor are the kernel module and a virtual machine monitor.
   a. True
   b. False
4. Virtual desktop infrastructure is the most widely used desktop virtualization technology when customer desktop applications are to operate from a virtualized data center.
   a. True
   b. False
5. Virtual provisioning simplifies the provisioning of storage to computer systems by making it only virtually available, independent of the physical storage capacity that has been provisioned.
   a. True
   b. False

Multiple Choice

1. Which of the following is a component of the hypervisor?
   a. Storage area network (SAN)
   b. Virtual machine monitor (VMM)
   c. Network-attached storage (NAS)
   d. Logical unit number (LUN)

2. Which technology establishes the remote connection between the user's device and the VDI server?
   a. Thin LUN
   b. Virtual machine monitor (VMM)
   c. Connection broker
   d. Kernel
3. Which of the following is a data storage virtualization technology that combines multiple disk drive components into a logical unit?
   a. Thin Pool
   b. VM
   c. RAID
   d. RDS
4. What is the purpose of virtual provisioning?
   a. Improve ease and speed
   b. Improve capacity utilization
   c. Improve performance
   d. All of the above
5. What is the most widely used desktop virtualization technology?
   a. Virtual desktop infrastructure (VDI)
   b. Virtual data center (VDC)
   c. Virtual machine (VMs)
   d. Remote desktop services (RDSs)

Short Essay Questions

1. Describe the difference between RDS and VDI.
2. How is storage capacity allocated to virtual machines from a SAN or NAS environment?
3. Briefly explain file-level storage.

Chapter 12

Analyzing Big Data

Introduction to Big Data

We are in a period that is characterized by an explosion of social media and its collection of diverse media types. Corporations and governments are increasingly interested in acquiring, storing, transmitting, and eventually analyzing the mix of telephone call data, email packets, short message texts, pictures, videos, audios, and all various forms of social media. The drive to address the sifting and mining of such varied information has triggered a growing push to systematically deal with these data on a collective and integrated basis. Pattern analysis software is attempting to scan mixed collections of these data for advertising, marketing, personnel scanning, and government profiling. This chapter explores the process behind *big data* and suggests approaches for formally analyzing it.

Among the multiple characteristics of *big data*, four stand out as its defining characteristics.

1. *A huge volume of data*. The tools that store, access, and analyze the data must be able to manage billions of lines of data at one time.
2. *A complexity of data types and structures*. Available and useful data are composed of an increasing volume of unstructured data. About 80%–90% of these data in which businesses, governments, and private organizations are interested are classified as unstructured data, which are sometimes described as part of the digital shadow or *data exhaust* of a process. Data exhaust is the data in social networks where the residue of communication, which is like the exhaust fumes of the user's communication process, might prove to be useful when collected and offered to a marketer wishing to discover and analyze collective and individual social behaviors.

3. *Constantly increasing speed in the creation and sharing of these new unstructured data*. Every day brings forth a new batch of data, merging with previous data, or obliterating already stored and analyzed data.
4. *Distinct analytical processes*. Due to the size and variety of structures exhibited by these new data, they cannot be easily and efficiently analyzed using only traditional methods applied against traditional databases. Where previously we might be storing records of text and numbers in a structured relational database with its fixed table structure, we now have *blobs* of unstructured but interesting data, which are difficult to acquire, parse, store, access, and then analyze.

These kinds of big data problems require new tools and technologies to store, manage, and benefit business. Due to the high volume and complexity of the data, the preferred approach for processing big data is employing parallel computing environments and experimenting with massively parallel processing (MPP). MPP enables simultaneous, parallel ingest, analysis, and storage allocation of data. The exploding cost and storage requirements of data have resulted in a movement to store the data in cloud data centers spread across hundreds of commodity processors and inexpensive storage devices. Moreover, since the majority of what are considered big data are unstructured or semistructured in nature, additional analytics techniques, extensive processing, and storage capability are required. Before exploring these techniques, let us examine the most prominent characteristic of big data—their structure.

The Structure of Big Data

Big data can come in multiple forms—from highly structured financial data, text files, multimedia files, and the variety of media types employed by social media. A high volume of data is the most consistent characteristic of big data. It features a variety of forms and formats with complex structural characteristics.

The Four Types of Data Structures

The most prominent and distinguishing characteristic of big data is their structure. There are four general types of data structures: structured, semistructured, quasi-structured, and unstructured data with 80%–90% of future data growth coming from the nonstructured data types (semi, quasi, and unstructured). The following are examples of what each of the four main types of the different data structures looks like.

1. Structured data: People tend to be most familiar with analyzing structured data since these are the numerical and alpha data we have processed in our corporate systems and stored in their databases.

2. Semistructured data have a discernable pattern and thus are susceptible to parsing. Much of the data on the web such as HTML or XML formatted data are semistructured.
3. Quasi-structured data are textual data with erratic formatting such as you might find in a clickstream string of bits.
4. Unstructured data, or unruly data, tend to have no fixed pattern. We find this with audio, image, and video streams, although we do segment them into pieces, which we then place in formatted packets for transmission.

In reality, these four different types of data can be mixed and used together in various ways. For instance, you may have a classic relational database management system (RDBMS) storing call log for a software support call center (such as what Cisco employs). In this case, you may have typical structured data such as date/time stamps, machine types, problem type, operating system type, and machine location, which may be entered by the support desk person from a pull-down menu or through a graphical user interface. In addition, you will likely have unstructured or semistructured data, such as free form call-log information, taken from an email ticket indicating the problem or an actual phone call description of a technical problem and a solution. The most salient information is often hidden in the unstructured text and voice. Until recently, most analysts would not be able to analyze the most common and highly structured data in this call-log history in an RDBMS, since the mining of the textual information in the past was both labor-intensive and difficult to automate.

Analyzing Big Data

Statisticians have been developing and employing routines to analyze data for over 100 years. Key routines include the following:

1. Categorization tools to separate data into known groupings with statistical comparisons among the groups. Among these are *K*-means clustering, which aims to partition *n* observations into *k* clusters in which each observation belongs to the cluster with the nearest mean, serving as a prototype of the cluster. This process is complex, but a number of heuristic models have been created that quickly converge on an optimum answer.
2. Application of association rules indicates relationships between the behavior of people and their similar characteristics, including what a particular person tends to like, purchase, watch, or visit.
3. Regression is the fundamental tool for determining relationships and answering the question of how much certain variables contribute to behavior. Regression might further be used to predict the lifetime value of a given customer or to predict the probability that a particular loan will incur default.

4. A naïve Bayesian classifier is a simple probabilistic classifier based on applying Bayes' probability theorem. This provides a model for statistical classifiers used in pattern recognition. Spam filtering is a prominent use case for naïve Bayesian classifiers. Classification might also be used to determine where in a catalog or in a store a product should be placed or whether an email is likely to be a spam.
5. Time series analysis is a predictive technique that companies, academicians, economists, financiers, and government workers have employed over the past century. Time series analysis might be used to determine the likely future price of a stock or what the sales volume will be next month, given a history of past data in those areas.
6. Text analysis is performed by parsing, classifying, and analyzing characteristics and associations of gathered information in order to glean new information beyond just the intended meaning of the text. This technique might be used to determine if a product review is a positive or a negative one.
7. Clustering is a popular method used to form homogeneous groups within a data set based upon their internal structure. Moreover, clustering is a method often used during exploratory analysis of the data by creating clusters of those with similar attributes. Using the notion of similarities, we can consider questions such as how to group documents by topic and how to perform customer segmentation into groups.

Analyzing the Four Data Types

Analyzing Structured Data

Let us look at how we might apply some of the previously identified common statistical analysis routines. Structured data are primarily numerical data frequently stored in databases. As a preliminary step, we routinely plot the data to observe underlying patterns and relationships. Then we gather variables that seem to have some relationship, either causative or predictive. We create a hypotheses, test it, and test our result by performing some form of regression. However, we do not just predict the outcome; we also determine how changes in individual variables affect the outcome and what portion of the effect can be attributed to each variable given the effects of the other variables. The outcomes can be continuous or discrete. When it is discrete, we are predicting the probability that the outcome will occur rather than the level of the outcome.

Some uses of regression include determining the expectation that customers might purchase items given that a set of other factors are in play. Or we might attempt to determine if a loan will default given a certain set of occurrences happening at a particular level. It can also deal with a multiclass problem of whether a voter or web purchaser will show up/purchase, or show up but not vote/not purchase. The logistic form of regression would be used to estimate probabilities of any of those actions occurring. Examples of such usage are binary classification problems containing the following types of likelihoods:

- Likelihood of responding to medical treatment or no response
- Likelihood of purchasing from a specific website or being unwilling to purchase from that site
- Likelihood that the Seahawks would win the 2014 Super Bowl

Naïve Bayesian classifiers are used to assign labels to objects employing the Bayes' algorithm. These labels have been predetermined, and the process is one of figuring out which label goes with which object and a probability that the classification has been correctly applied. Naïve Bayesian classifiers are among the most successful algorithms for classifying text documents, which will be discussed in the following section on unstructured data. They are also employed to filter spam email and detect fraud in insurance claims.

Clustering and decision tree analysis are methods often used in the early exploratory analysis of the data. Clustering is a popular method used to form homogeneous groups of like subjects within a data set based on their internal structure. There are no predictions of any values done with clustering, just finding the similarity among the data and then grouping them into like clusters under the notion of similarities. The definition of similarity is specific to the problem domain or area that needs to be examined to solve a problem. We define similarity as those data with the same topic tag, or customers who can be profiled into the same age group, income, or gender, or who fall into a common purchase pattern. Some examples of questions that decision tree analysis can help answer are as follows:

- How do I group these documents by topic?
- How do I perform customer segmentation to create targeted or special marketing campaigns?

There are many clustering techniques, but one of the most popular clustering methods is known as *K*-means clustering. With this technique, the data measurements of an attribute of the clustered data will have values for the measurement rather close to each other. The distance between the points within a cluster (small for similarity) should always be lower than the distance between the points in a different cluster. In each cluster, we end up with a tight homogeneous group of data points that are far apart from those data points in a different cluster.

Time series techniques allow forecasting of the data by isolating and projecting the patterns (trend and seasonality) of past data into the future. A commonly employed time series analysis methodology is the Box–Jenkins method developed by George Box at the University of Wisconsin. It is formally known as ARIMA or autoregressive integrated moving average forecasting. The Box–Jenkins method projects long-term patterns based upon analysis of the series itself at particular influential points of the past. These influential points are plotted and smoothed to remove peaks and valleys in the data by applying 1 and 12-month differencing. This technique may be applied to quickly determine forecasts that are uncomplicated in form or that involve a number of economic variables. In either case, use of this technique enables efficient utilization of other predictive information contained in the data. It offers assurance of obtaining the highest forecasting accuracy possible in terms of the variables on which the forecast is based.

Many of these categories of techniques overlap with the others and with the type of problem they can address. Questions such as "How do I group these documents?," "Is this e-mail spam?," or "Is this a positive product review?" are types of questions that can be answered by means of classifying the data into similar groups. However, these questions can also be considered a text analysis problem. Text analysis is another process of representing, manipulating, predicting, and learning from text by classifying into like clusters.

Analyzing Unstructured and Semistructured Data

Text analysis is essentially the processing and representation of data that are in text form for the purpose of analyzing for patterns, understanding new conclusions, and creating models from that information that you can employ against newly acquired data. The main challenge in text analysis is the problem of high dimensionality. When analyzing a document, every possible word in the document represents a dimension not only offering possibilities but also adding to the complexity. The process of text analysis is composed of three important steps: parsing, tagging, and text mining.

1. *Parsing.* Parsing starts with monitoring or acquiring input data from sources and then parsing the sentences of those text inputs. It is the process of breaking up each sentence into its component parts of speech and then trying to show relationships contained in the syntax of the sentence. In this context, we are talking about parsing HTML pages, RSS feeds, blogs, basic Internet HTTP messages, and all the various forms of HTML-based web data. In doing this, we need to impose enough structure upon the bulk raw text so that we can find the parts of that raw text in which we are interested. This might be the content of published pieces (either formally published or on an Internet site/blog), their titles, description, and an indicator of by whom and where it was reviewed and when it was posted. Parsing requires knowing the format of the data source. Sometimes the grammar is relatively standard such as HTML or RSS text. Other times, it may not be quite as standard when it comes from web logs or blogs.

   Once we have extracted, parsed, and categorized our chosen text, then we have our content. Next, we want to determine if this found and identified content is actually of interest to us. *Regular expressions* are a popular technique used for finding words, strings, or particular patterns in the text. The basic approach is to employ a file of standard expressions, run it against the data collection, and extract the segments that seem to match that expression file. We would decide beforehand what we would consider a successful match. For example, we might decide that if 75% of the words in our text string match an expression in the regular expression file, we consider it a successful match. With regular expressions we can also adjust for capitalization or lack of capitalization, common misspellings, common abbreviations, and common web shortcut words.

2. *Tagging.* Tagging itself is a complex problem, and to properly tag a document is not always a straightforward process. Some of the methods employed are as simple as counting the number of occurrences of a product name to determining networks of relationships between the products, suppliers, customers, and needs, which might require sophisticated methods.

   There are rules you can employ to determine how to sort a document in a given context. If an item is mentioned in the title, then the document is likely to be about that item, although mentions of another item in the body of the text may or may not be relevant to the tag. So tagging that document may require more than just simple analysis.

   In another context, one might conclude that a tweet that mentions a product is probably about that product, whereas a review may compare many products and not be specifically about any one of them. Thus, the source and context of the document are meaningful to the tagging process. However, more frequent mentions of the product in the document rather than a single mention are likely a clue to the importance of that product in that document.
3. *Mining and understanding the meaning of the text.* The third step is understanding the content itself. Instead of treating the text as a set of tokens or keywords, in this step, we attempt to derive meaningful insights into the data pertaining to the domain of knowledge, business processes, or the problem we are trying to solve.

Once we have parsed all our data inputs, collected the phrases and words, and categorized and tagged them, we are ready to represent what we have collected in a structured format for downstream analysis. The most common representation of the structure of text is known as the *bag of words*. The bag of words is a unit of related words (called a vector in big data and mathematical circles) with there being one dimension for every unique term in the text space and a notation for term frequency. Term frequency is the number of times a term occurs in a set of words or word vector.

The word vector can be quite small or very large and of a high dimensionality. We invariably end up with a significant number of unique words in a document. Bag of words is a common representation, and it is well suited for search and classification processes. We count the occurrences of each of the words in the parsed text and the number of times the word is repeated. We then store that word count as a part of the vector representation.

In order to reduce the dimensionality of the word vector or bag of words, we do not include all words in the English language. Normally we ignore some common connecting words such as *the*, *a*, *an*, etc. We also tend to avoid pronouns in the term. The word *vector* we have categorized must be managed and stored so that it only contains words that are essential for our analysis. This is done based on the context of the acquired original document and its parsed word units. In a completely unstructured document, special techniques such as parts of speech tagging are used as part of the parsing process to minimize the number of unnecessary words to be stored.

In the case of sentiment analysis, algorithms can be created based on data origins, such as sites that have quantitative ratings for specific products. But the idiosyncratic terminology of the reviewing website, the community of interest, or the product category may influence or cloud the general usage of that information. For example, classifiers built from reviews are unlikely to work on tweets or blog comments due to the special terminology that might be employed in tweeting or blogging, the abbreviations and code words used, and the special interest of a particular user community. Each site must be judged based upon the universality of the extracted information and sentiments embodied in that information.

One might precluster the documents and then assign labels based on whether or not sampled documents from a cluster are positive or negative. But if the cluster is not built specifically on a particular sentiment, it may not partition on that sentiment after the fact.

Other procedures you can do to track sentiment besides attempting classification are mixing computer preliminary analysis with human decision making. You can track the frequency with which certain words appear in reviews of products, and then let a human decide if the overall trend looks positive or negative. This can be laborious with large volumes of data but can be appropriate in specific cases where the amount of text a human evaluates is relatively small.

One difficulty with the whole text analysis process is that each day brings a whole new set of data waiting to be extracted, parsed, categorized, and then searched. It is an unending process. So we are faced with a growing volume of information, with varying frequencies in which those data change. Therefore, we must think of big data analysis as a dynamic and unending proposition. All of our categories, tags, indexes, and metrics must be updated continuously as new information daily becomes available.

The Corpus or Body of Work

After creating a representation of each piece of text that results from the individual parse, tag, mine, and understand processes, we then need to represent all the information in a collected body of work called the *corpus*. This is done by creating a reverse index, which provides a way of keeping track of the list of all documents that contain a specific feature, the frequency of those features, and every possible feature that might exist. Other corpus metrics such as volume text components and the corpus-wide term frequency, which specifies how the terms are distributed across the corpus, help with the downstream analysis and processes of classification and searching. Search algorithms also employ inverse document frequency.

A fact that many people do not think about is that documents are often only relevant in the context of the whole corpus, or a specific collection of documents. Sometimes this is obvious, for example, employing search and retrieval. It is less obvious in the case of classification like when performing spam filtering or sentiment analysis. But even in that case, the chosen classifier has been trained on a specific set of documents, and the underlying assumption of all classifiers is that they will be deployed on a population that is similar to the population on which they were trained.

Reviewing Results

In the end of the mining and understanding phase, when we have our corpus tagged, classified, analyzed for sentiment, and stored, we need to be able to call it up on demand by a specific indicator, tag, or named document of interest. Searching is typically performed by means of issuing a query command. The query process might specify calling up of documents from a particular site or calling up reviews in a specific data range. This is a search problem of finding the document that meets the search criteria.

Afterward we must focus on the quality of our search results. This is the process of determining if the results you receive are indeed the ones you wanted. Relevance, precision, and recall are the metrics used to determine the quality of such search results.

Since many stored documents may meet the search criteria, a ranking system must be created that ranks the search results based on their relevance to the requestor and provides users the most relevant documents ahead of those that score lower on the scale of relevance. Some examples that are commonly employed are

- Authoritativeness of the source
- Most recent documents
- A record of how often a document has been retrieved by other users

By these relevancy rankings, we can provide some measure as to whether this is the document that the user wanted.

Summary

With the available variety of media forms, technology appliances, transmission capability, and social websites, people are now able to communicate with each other no matter the time of day, location, and even the availability of either parties. This plethora of information contained in these communications stimulates interested parties, organizations, and governments to examine these data and to come to conclusions about people's characteristics, behaviors, and associations.

Much can be gleaned by applying sophisticated parsing, mining, translation, classification, clustering, and other analytic tools to these data. With the growing availability of massive parallel processing engines and large, intelligent storage arrays, interested parties are finding more and more useful information that they can use to predict and manipulate people's behavior. Such is the promise and threat of big data analytics.

Review Questions

True/False

1. Nonstructured data types will make up 80%–90% of future data growth.
   a. True
   b. False
2. A naïve Bayesian classifier is a simple probabilistic classifier based on applying Bayes' probability theorem with strong independence assumptions.
   a. True
   b. False
3. The process of parsing, tagging, and mining characteristics and association of gathered information is part of time series analysis.
   a. True
   b. False
4. The current majority of big data is considered *structured.*
   a. True
   b. False
5. Statisticians have been developing and employing routines to analyze structured data for over 100 years.
   a. True
   b. False

Multiple Choice

1. What is the most prominent and distinguishing characteristic of big data?
   a. Their volume
   b. Their structure
   c. Their variety
   d. All of the above

2. Which of the following is not one of the four general data types?
    a. Structured
    b. Semistructured
    c. Quad-structured
    d. Unstructured
3. What is generally the first step in analyzing unstructured and semistructured data?
    a. Compiling
    b. Tagging
    c. Parsing
    d. Categorizing
4. Which of the following is not an analytical method used with big data?
    a. Regression
    b. Text series analysis
    c. Categorization
    d. Time series analysis
5. What are the metrics used to determine the quality of search results?
    a. Relevance
    b. Precision
    c. Recall
    d. All of the above

Chapter 13

The Cloud and Cloud Computing

Introduction

The world is awash with data that are overwhelming our ability to store and analyze them. The enormous cost of constructing, operating, maintaining, and expanding data centers has led to the implementation of cheaper, more flexible centers that can provide on-demand services to meet user requirements. Furthermore, the increased access to data through enormous growth in social media and ever-present use of portable devices has further pressured data center providers to implement a new model of processing, storing, networking, and even using *desktop* applications on less expensive, commodity personal computing servers. This new model of operation is termed *cloud computing*.

For those who are intimately involved with the cloud and its offerings, it is viewed as either the great *liberator* or the great *destroyer*. As a liberator, it could free us all from the persistent updating of our computers and liberate us with always having information available anytime, anywhere, and with any device; plus it could be a cheaper one-time purchase. As a destroyer, it could change ICT focus immensely, condense all our information into one location (to be destroyed, stolen, etc.), and ruin privacy. This chapter will explore the cloud and discuss where it has come from and where it is going.

The History of Cloud Computing

The concept of cloud computing is not a new phenomenon. Some people trace the usage of the term to 2006 when Amazon and Google came to use it to describe web

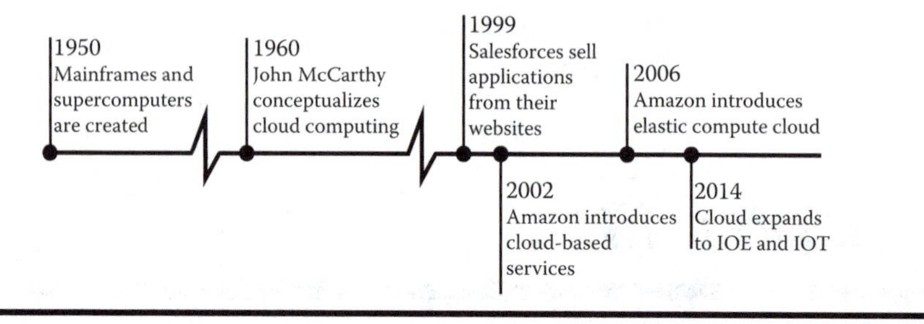

Exhibit 13.1 Cloud computing timeline.

access of computers, software, and files. Others date it back further to 1996 where Compaq Computer described business happening on the Internet. In actuality, the concept of the cloud, sharing resources, or cloud computing dates back far longer to the 1950s (see Exhibit 13.1). Then IBM's mainframe computers dominated the computing world. These mainframes were room-sized, expensive, and difficult to maintain. In order to cut the operating cost and allow more user access, they created a virtual machine (VM) interface, which allowed multiple users to share access and computing power, therefore providing a better return on investment for companies. This idea for resource sharing allowed users to be connected to a central computer and, through it, users could share data and the cost of the computing. This was the seed of the idea behind cloud computing and accessing storage via the Internet.

With the twenty-first century came a lot of innovations with the cloud. In 2008, Eucalyptus became the first application program interface (API) compatible platform for private clouds. Efforts now are focused on providing some type of quality of service (QoS) for users. During 2009, the Gartner Research Company recognized cloud computings' ability to change the relationship between an ICT user and a provider—a groundbreaking movement. The year 2010 brought out the National Aeronautics and Space Administration's and Rackspace's efforts toward an open-source cloud software product. In 2011, IBM launched its SmartCloud Framework for businesses. In 2012, Oracle announced the Oracle Cloud. Since then, not a year has gone by without some major announcement on the cloud, a pivotal technology release, or a new business model.

What Is the Cloud?

There are many different definitions of the cloud. Gartner, an ICT research firm, defines it as "a style of computing in which scalable and elastic IT enabled capabilities are delivered as a service using Internet technologies." The National Institute of Standards and Technology (NIST) defines it as "A model for enabling ubiquitous, convenient, on-demand network access to a shared pool of configurable computing resources (networks, servers, storage, applications and services) that can be rapidly provisioned and released with minimal management effort or service provider interaction." Of all the definitions, four elements always come into play:

1. *On-demand self-service*: Customers of cloud computing can request services on-demand and generally arrange those services as they need them. Traditionally, users have to install software packages, such as Microsoft Word or Microsoft PowerPoint, in order to use them. If the user is away from the computer where the software is installed, this software is no longer available for usage. However, now much of the required software functionality used can be accessed over the Internet. Free Internet versions of common software allow users to access programs, edit files, and store resources from any device that has Internet connection, thus eliminating the need to have access to a particular office-based computer device.
2. *Shared resources*: Cloud computing distributes computing power between multiple users. Traditional computing is like a personal automobile. It probably sits in a driveway or a parking lot majority of the time. Cloud computing is like a public transportation system in which users share a common infrastructure. A cloud computing computer utilizes its resources more efficiently because the hardware is used to its full capacity: when you are not using the computing power, someone else is.
3. *Rapid elasticity*: Cloud-based resources, including both processing and storage, can be automatically and dynamically allocated, expanded, and contracted quickly and efficiently without interruption of service. When users experience large fluctuations in their required capacity, the cloud provider can temporarily increase the number of web and application servers for the duration of a specific task or for a specified period of time, and then contract when the demand subsides.
4. *Measured service*: Cloud computing service providers keep track of customer usage with a metered service system. They then provide billing and chargeback information for the cloud resource used by the consumer. The metering software continuously monitors CPU time, bandwidth, and the storage capacity used and regularly provides reports concerning that usage to the consumer. The customer only pays for the actual capacity used and not for any standby capacity that might be reserved in case of temporary need. When organizations need to rapidly expand their business and the computing capacity to support those increased operations, cloud providers quickly accommodate such requirements without the need to raise capital and purchase additional equipment. The customer merely needs to request expanded facilities and configure them for their usage.

Virtualization

Providing these cloud computing services requires the virtualization of a network device. Virtualization is the technique of separating the physical hardware from the

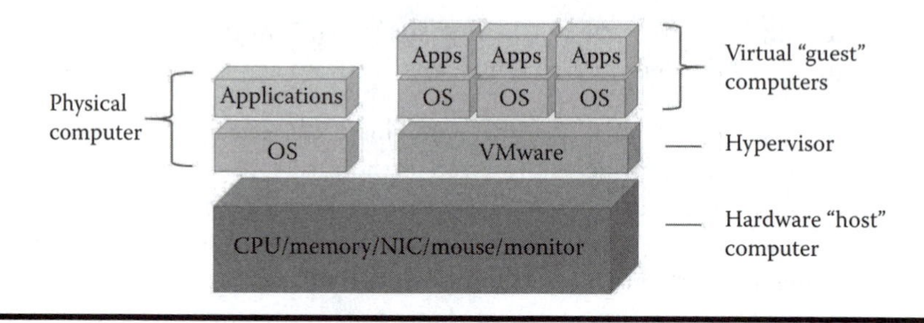

Exhibit 13.2 Virtualization model.

operating system (OS). This then enables multiple virtual OSs to be added to and run at the same time on a single machine.

The process of virtualization begins with a host device such as a computer. Virtualization software, called a hypervisor, is installed on the host machine, which creates and manages one or more guest virtual OSs. The hypervisor sits between the OSs and the computer hardware, allowing multiple applications and OSs' software to execute through the hardware of a specific physical computer that hosts them (see Exhibit 13.2).

After virtualization, cloud computing can be used to extract standard desktop computing and application operating services on that desktop to a virtual data center (VDC) where portable devices can access them. This process allows users with their smart mobile phones and tablets to have the power of an office computer while maintaining the small size, mobility, and portability of a mobile device. Operating these VDCs provides flexibility, improved resource utilization, and ease of management compared to the operation of traditional data centers, enabling them to operate more effectively. Virtualization was further covered in Chapter 11.

Delivery Models for Cloud Computing

Cloud computing services are supplied along three basic models to which unique names have been identified.

1. *Infrastructure-as-a-Service (IaaS)* is a base layer cloud computing service and often serves as the foundation for the other two offerings: platform as a service (PaaS) and software as a service (SaaS). The cloud infrastructure, consisting of servers, routers, storage, and networking components, is provided by an IaaS cloud infrastructure provider. The customer hires these resources as a service based on their needs and pays only for the usage. The customer is able to deploy and run his/her own OSs and application software. In this situation where only the infrastructure is provided (on an as needed and used basis), correct estimation of the scaling of the required

resources and elasticity of their moment-to-moment usage are the responsibilities of the consumer, not the provider. Thus, IaaS is a bare-bones cloud data center service where the infrastructure is provided, but the customer must configure the required resources (server, storage). Major responsibility is undertaken by the customer with this level of service in order to meet cloud service needs.

2. *Platform-as-a-Service (PaaS)* can be thought of as both an application development environment as well as an application operation environment, which is offered as a service by the cloud provider. With this delivery model, the user controls or creates the application but uses the clouds' underlying infrastructure and OS. Customers can either deploy applications of their own creation on the infrastructure offered by the cloud provider or purchase applications from the provider. This service saves the customer from the bother of upgrade pains as well as the complexity of evaluating, buying, configuring, and managing the software and hardware infrastructures. Elasticity and scalability are guaranteed by the PaaS cloud provider when applications are run over the PaaS provider's software platform.
3. *Software-as-a-Service (SaaS)* is the most complete service offering of the cloud computing stack. It offers the consumer the capability to use the cloud service provider's applications running on the cloud service provider's OS and infrastructure. In a SaaS model, the applications such as customer billing, accounts receivable, customer relationship management (CRM), sales management, email, and instant messaging (IM) can be prebuilt and offered as a commodity application service by the cloud service provider. The customer accesses these provided applications from a web browser, a thin client interface, or a program interface (App). The customer only uses the applications he/she needs and pays a subscription fee for that usage. The cloud service provider hosts and manages the required infrastructure, the management and control application tools to support these services, and the set of prebuilt and packaged application suites of services that the customer requires. SaaS reduces the need for infrastructure since storage and computing resources are provided to the user remotely over the network. The SaaS providers can perform many of the manual update tasks for the user automatically, significantly reducing the amount of customer workload.

Deployment Models for Cloud Computing

Three broad deployment models have been used to provide cloud computing services. These deployment models offer these just described three levels of cloud service offerings—computer PaaS, SaaS, or IaaS—in a specialized fashion and may even offer all three service levels simultaneously by a particular cloud data center facility.

Public Cloud Deployment

IT resources are made available to the general public or organizations and are owned by the cloud service provider. The cloud services are accessible to everyone via standard Internet connections. In a public cloud, a service provider makes IT resources, such as applications, storage capacity, or server compute cycles, available to any consumer. This model can be thought of as an *on-demand* and a *pay-as-you-go* environment where there are no on-site infrastructure or management requirements. However, these benefits come with certain risks for organizations: the customer has no control over the resources in the cloud data center, over the security of confidential data, and over network performance issues, and has little control over interoperability.

Private Cloud Deployment

The cloud services are operated solely for one organization and are not shared with other organizations. This cloud model offers the greatest level of security and control but significantly reduces the ability to share resources across a number of companies' computing requirements. There are two variations to a private cloud:

a. *On-premise private clouds*: On-premise private clouds, also known as internal clouds, are hosted by an organization within their own data centers. This model provides a more standardized process and the customary protection but is limited in terms of size and scalability. Organizations would also need to incur the capital and operational costs for the physical resources. This is best suited for applications that require complete customer control and configurability of the infrastructure and security.
b. *Externally hosted private clouds*: This type of private cloud environment is hosted externally by a cloud provider for a specific customer organization with a full guarantee of privacy or confidentiality. This is best suited for organizations that do not prefer a public cloud due to data privacy/security concerns, but also wish to avoid the burden of financing and operating their own data centers.

Hybrid Cloud Deployment

Customers can combine private and public cloud services to maintain service levels in the face of rapid workload fluctuations. For example, organizations may use their computing resources within a private cloud computing center for normal usage but access the public cloud for less risky, high/peak load requirements. This ensures that a sudden increase in computing requirements is handled gracefully. An organization might use a public cloud service for archiving certain data but might continue to maintain in-house storage for important operational customer data. Ideally, the hybrid approach allows a business to take advantage of the scalability and cost-effectiveness that a public cloud computing environment offers without exposing mission-critical applications and data to third-party vulnerabilities.

Community Cloud Deployment

This is an offshoot of the private cloud where a multitude of institutions or organizations get together to create a private cloud. Community cloud might be managed by the organizations or by a third party with the costs spread over fewer users than a public cloud. Although the community cloud option is more expensive than a public cloud offering, it offers a potential of a higher level of privacy, security, and policy compliance as well as access to a larger pool of resources than would be available in a private cloud offering.

Risks of the Cloud

With cloud computing comes the security risks of putting company and personal data in the cloud Exhibit 13.3 illustrates five common cloud computing risks. The security risks and ramifications of those risks are numerous and should be fully understood before implementing cloud computing in your home or organization. Below are five broad categories of risks with cloud computing.

Multitenancy: When you live in an apartment building, your neighbor's actions affect you. Noise, fire, and theft can all affect you. In a multitenancy cloud computing environment, multiple users remotely utilize the same OS on a piece of hardware. Shared resources (CPU, storage, memory) are exposed to many potential security breach points.

Virtualization security: In a virtualized cloud computing environment, multiple users remotely utilize different OSs on a single piece of hardware. Points of potential security breaches exist in four main environments: server host only, guest to guest, host to guest, and guest to host. Data theft, attacks, and viruses can be introduced at each of these layers of vulnerability.

Access: Insufficient authentication requirements and authorization levels are a significant security issue in a cloud computing environment. Knowledge of how

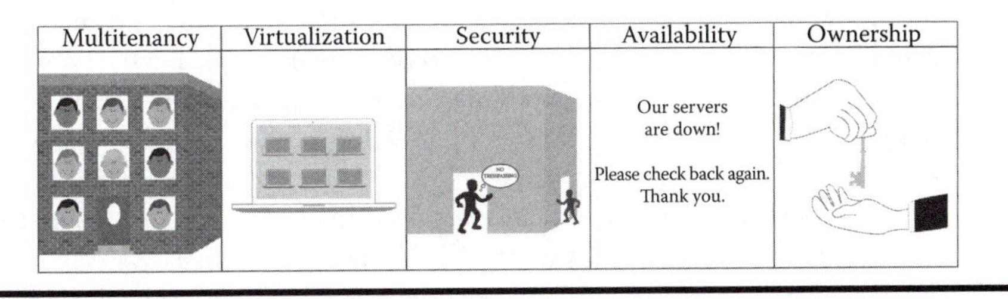

Exhibit 13.3 Risks of the cloud.

many internal people have privileged accounts, how private keys are shared by the provider among tenants, and how data are disposed of is vital to securing a cloud environment.

Availability: Availability is a key selling point of the cloud, but there are many factors that could lead to availability failure. It is important to know the cloud providers fault tolerance, uptime, and redundancy capabilities. Provider contracts should address responsibility in the event of data loss. Even large reputable providers can lose access for hours or even days. Another issue to consider is wireless area network (WAN) availability. Although the cloud provider and your site could be fully operational, the WAN provider in between (DSL, cable, wireless, metro Ethernet) could be down and therefore prevent access to the cloud-based data.

Ownership: Who really owns the data? The customer or the cloud provider? If the cloud provider goes out of business, what happens to the data? Knowledge of and legally binding agreements on these points are vital to the long-term security of data stored on the cloud.

The process of preventing cloud-based security risks is new to many organizations. However, with adequate service level agreements and a track record of performance by cloud vendors, customers can assure themselves of a near-term QoSs. The advantages to be reaped from cloud computing should be in the plan of companies while they remain ever vigilant that change will occur in the future.

Future Predictions

So what will the cloud be like in the future? No one can definitively say, but there are streams of logic that all point to some commonalities.

1. *The cloud will grow*: Gartner Research expects adoption of cloud-based services to grow to $250 billion in the next few years, this growth being predominantly by companies that want to develop, market, and sell products and manage their internal operations. SaaS will especially grow high as companies try to sell all of their products *as a service* via the cloud.
2. *The hybrid delivery model will expand*: Trying to put all services in the public cloud will not work well due to security concerns, complexity of services, and management of outsourced processes. Putting all services into a private cloud could be more expensive than it needs to be. A hybrid cloud delivery model provides the best of both worlds. What is required is a strategic blending of both delivery models with the appropriate selection for each. Today's world has many companies putting email, CRM, and hosting services into the public cloud while maintaining financial services, inventory, and personnel data in the private cloud. As time progresses, this mix will change.
3. *Increased innovation*: As the cloud creates new services and products, innovation and competition will increase. This will cause increased development

of cloud-based services, especially SaaS. Application developers will become more important.

4. *Strategic differentiator*: The cloud and the services offered will be a differentiator for companies. This will become especially important, as more verticals and companies become software dependent to provide their services. As one example, you could have your blood pressure monitored and put into the cloud so your doctor can read it in real time.
5. *Competition will increase*: Initially, Amazon ruled the cloud world with others following their lead. Other major players are becoming increasingly cognizant of the value that the cloud can offer and are directly targeting the competition. It is unsure how competition will shape cloud services. Perhaps it will be similar to the airline industry where prices will decline, but service, service reach, and comfort will also decline.

Summary

Cloud computing offers companies an alternative to constructing, staffing, operating, growing, and maintaining their own data centers for processing their corporate records, supporting business functions and personnel, and offering services to the company's customers. By means of cloud service offerings, companies can now offload the costs and responsibilities of data center operations while paying for the usage of cloud data center services on a pay-as-you-go basis.

In this chapter, we introduced the cloud computing concept and defined its major characteristics: on-demand self-services, pooling of resources, scalability, and measured services, as well as the role of virtualization in supporting those services.

We then moved on to explaining how it is delivered. IaaS offers the hardware to you virtually for your use. PaaS offers the hardware platform and typically an OS for users to develop and/or run their applications. SaaS offers applications directly to the user with no modification.

The cloud can be deployed to you in many variations. In a private cloud, you have access only if you are a member of that corporation, organization, or group. In a public cloud, everybody who has Internet access can avail themselves of the delivery models available. A hybrid cloud is a combination of both the private and public cloud and is the fastest growing model of deployment. A community cloud is a private cloud that is made available to a group of people from different organizations that sign up for its use.

We next studied the risks of cloud computing. The risks mostly involved security, access, ownership, and availability. Unwise consumers or businesses can be very vulnerable if they are unaware of potential losses.

We finally ended with some future predictions concerning the web. There are a lot of money, movement, and success at stake here, and a lot of attention is being paid to it. We can be certain that the future of cloud computing will affect the workers of tomorrow and the skill sets needed for job success. By understanding the pros, cons, and working parts of the cloud, we will be able to better utilize its resources.

Review Questions

True/False

1. On-premise private clouds, also known as internal clouds, are hosted by an organization within their own data centers.
   a. True
   b. False
2. Rapid elasticity is a cloud characteristic in which the customer only pays for the services and storage capacity that he/she uses and not for any standby capacity that might be reserved in case of temporary need.
   a. True
   b. False
3. Cloud-based resources, including both processing and storage, can be automatically and dynamically allocated, expanded, and contracted quickly and efficiently without interruption of service.
   a. True
   b. False
4. The concept of the cloud, sharing resources, or cloud computing dates back as far as 1996.
   a. True
   b. False
5. Virtualization is the technique of separating the physical hardware from the operating system, enabling multiple virtual operating systems to be added and run at the same time on a single machine.
   a. True
   b. False

Multiple Choice

1. Which of the following is not an advantage of cloud computing?
   a. Flexible scaling
   b. High availability
   c. More energy consumption
   d. Reduced IT costs

2. Which of the following is the base cloud computing service offering?
   a. IaaS
   b. PaaS
   c. SaaS
   d. BaaS
3. The cloud infrastructure that is operated solely for one organization and is not shared with other organizations is what type of cloud service?
   a. Public cloud
   b. Secure cloud
   c. Private cloud
   d. Exclusive cloud
4. In a SaaS model, which of the following can be prebuilt and offered as a commodity application service by the cloud service provider?
   a. Customer relationship management (CRM)
   b. Instant messaging (IM)
   c. Accounts receivable (AR)
   d. All of the above
5. What are the risks of cloud computing?
   a. Availability
   b. Vandalism
   c. Multitenancy
   d. Both a and c
   e. All of the above

Short Answer Questions

1. What is Garter's definition of the cloud?
2. Name the four essential characteristics of cloud computing.
3. What is the benefit of a hybrid cloud deployment approach?

Chapter 14

Video Basics

We have discussed two forms of electronic communication: voice and data. There is a third type of communication that we will discuss in this chapter: video communication. Pictures, moving or still, are powerful forms of information and, as the saying goes, "a picture is worth a thousand words." Video can be powerful, but it is also very complex to create and to transmit. This chapter looks at the basics of video and how images are created and transmitted.

The Business and Human Factors

The ability to capture and transmit video images has led to the birth of multibillion dollar industries, such as television, that allow viewers to fantasize and "be there" without leaving the comfort of their own home.

A simpler, less robust, and more down-to-earth flavor of video in the business world is videoconferencing (VC) (Exhibit 14.1), which allows people to see and hear each other over great distances, reducing the amount of travel time and money spent. In a world of international companies and markets, VC is a blessing.

Similar to all communication systems, VC starts with basic elements: the sender, its transmission media, and the receiving station. VC is full duplex; the sender and the receiver are constantly interchanging information; thus, at any one point in time, the person who is sending voice and video information to a distant receiver is also receiving voice and video information. This provides the "next best thing to actually being there."

From the basic concepts of VC, there are three essential categories or qualities of VC systems: high-quality, medium-quality, and low-quality systems.

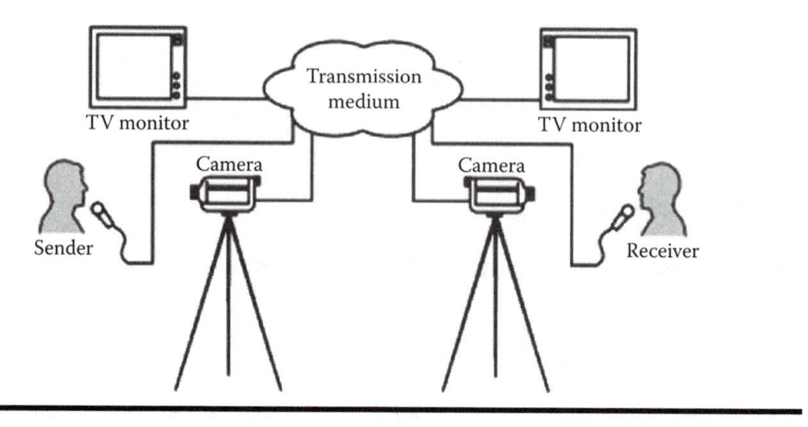

Exhibit 14.1 Basics of videoconferencing.

High-Quality VC

In high-quality VC systems (Exhibit 14.2), the transmission medium is called full frame, and the images seen are the same quality we see on our television sets. Video images are typically recorded in a professional studio with high-priced video cameras and microphones. From there, the video is sent via a high-bandwidth transmission medium to the receiving site. The physical transmission medium is typically satellite or fiber optics capable of carrying large amounts of data at very high speeds.

At the receiving end, the video information is transferred back into images and sound on high-resolution TV monitors and high-quality speaker systems.

Because of the amount of data transferred and the speed necessary, high-quality VC is used very little and only by large firms with very deep pockets.

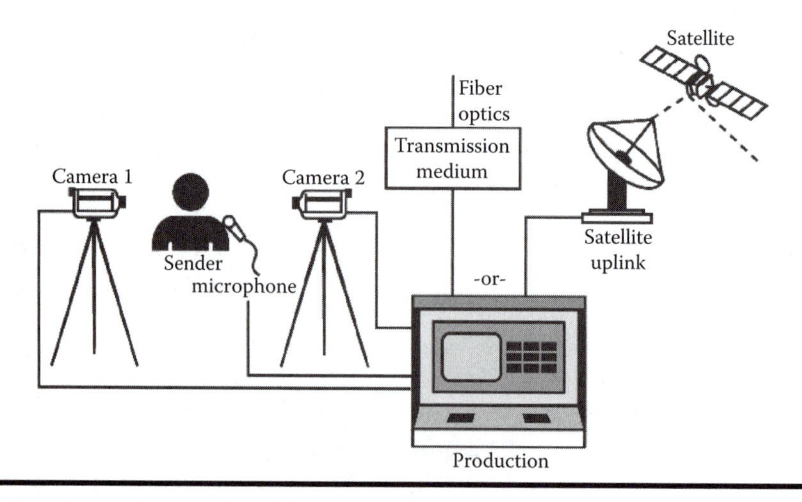

Exhibit 14.2 High-quality videoconferencing.

VC can connect multiple sites, as well as just a pair of sites. For example, a central sending site can send its transmissions to all sales branches in the United States, and the central site will be able to see and hear all the remote sites. This is far more effective than a telephone call for getting large, dispersed groups together in a simple, affordable way.

Be aware that there are firms that specialize in VC, and one may rent facilities to do this, which means you can get the highest quality for a one-time videoconference at a reasonable cost.

Medium-Quality VC

A more common type of VC is medium-quality VC (Exhibit 14.3). The compromise here is trading quality of sound and picture for cost. The video is of lower quality than commercial TV quality, with sound that is a bit harsh, but the cost is much lower than with high-quality VC systems. In a medium-quality VC system, lower-priced cameras and microphones are used at the sender and receiver sites, and the video information is compressed and limited to reduce the amount of bandwidth that is needed to be sent over the transmission medium.

Typically, the transmission medium used is the Internet or a specialized phone line called the integrated services digital network (ISDN). By using a more commonly available transmission medium, the cost of VC becomes affordable by most companies and people.

Many companies, such as PictureTel, make equipment for medium-quality VC, which has become fairly common in the workplace. The equipment comes on a rolling cart with the cameras, monitors, and compression equipment, making it fairly easy to use.

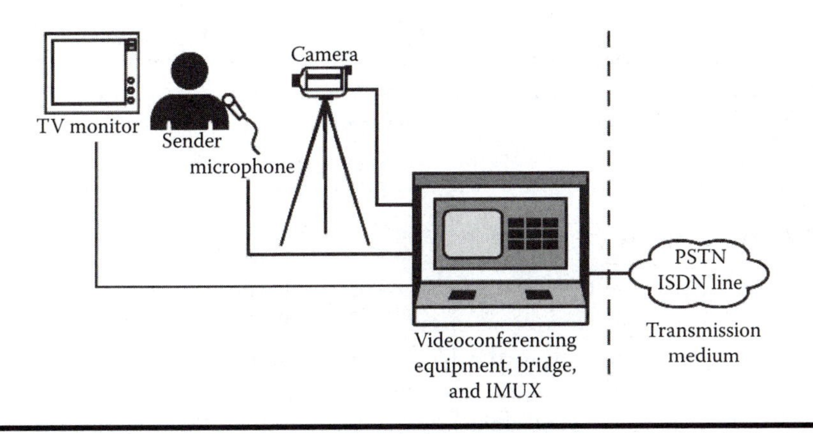

Exhibit 14.3 Medium-quality videoconferencing.

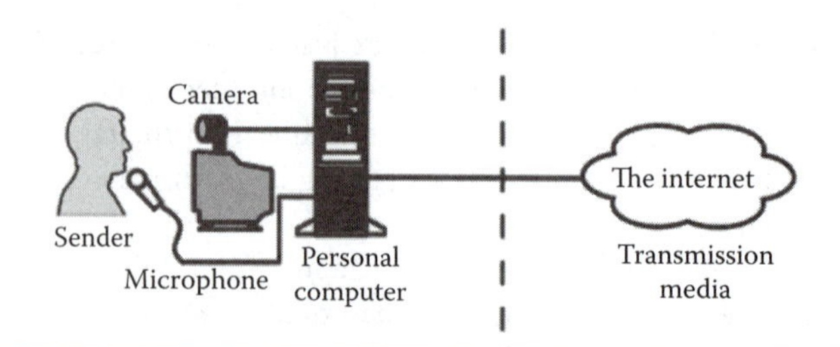

Exhibit 14.4 Low-quality videoconferencing.

Low-Quality VC

Low-quality VC systems (Exhibit 14.4) are becoming popular these days because they require minimal equipment at a minimal cost. Again, the trade-off for low-cost VC is quality of video and sound. In a low-cost VC system, the images are erratic and delayed, with minimal color and sound quality. The essence of the message is there, but when comparing the sound and video to television, the user is disappointed at the lack of quality.

Low-cost VC systems function by using commonly available equipment in a home or office. A low-cost camera and microphone (available for less than $200) are added to the PC and its Internet connection (LAN and WAN), providing a low-cost system. Computers with the Windows operating system are equipped with NetMeeting, a Microsoft package that enables VC from the PC. Add a camera and a microphone to your PC, and you are ready to videoconference with friends and business associates.

Those of us who have traveled a full day to attend a one-hour meeting know that VC can save large amounts of time and money, two precious elements in our personal and business lives. Is VC the ultimate solution? No, but it is the "next best thing to being there" and offers tremendous advantages.

The selection of a VC system, whether high, medium, or low quality, depends on the funds available and the quality needed. We would all like the highest quality, but often we do not need it nor can we afford it.

The Technical Factors

We must first start off with a human principle: persistence of vision. Persistence of vision is the principle at work when video information frames presented in rapid succession give the illusion to the TV viewer of smooth motion. Each of us has seen this in our childhood in the form of cartoons—still pictures photographed in rapid succession, making inanimate characters come alive. Maybe we have actually produced this effect by drawing stick figures on a pad and flipping through the pages. We even see this effect with light bulbs: light bulbs go on and off 60 times per second, but our eyes and brains see only a continuous light with no flicker.

Images appear free of flicker and look smooth when the presentation rates are above 40 times per second. Motion pictures use a shutter rate of 45 images per second, and television images use a rate of 60 images per second to impart continuity to the action. The result is that the human eye and the brain see images that are seamless.

At this point, we must also differentiate film pictures from video pictures. With film pictures, as seen at the movie theater, the complete picture is projected in front of us at a moving rate of 45 images per second. On the other hand, video pictures (in the form of TV images) are scanned, line by line, at a high rate of speed (6 million bits per second) to produce a visual image rate at 60 images per second.

Image Scanning

Scanning starts by taking light, bouncing it off a subject, and focusing it on a target (Exhibit 14.5). Image scanning (Exhibit 14.6) consists of *looking* in horizontal passes across the target image and measuring the instantaneous amplitude that represents the reflected light energy. Each pass across the image by the scanning element is called a *scan line*. The video camera contains an image tube or target that automatically scans the scene and generates a corresponding electric output, which replicates the image that the camera sees in electronic form. Similarly, the video receiver reproduces the scanned image on the viewing screen. The process of reproduction is done by a spot that scans across the viewing screen and varies its intensity in accordance with the signal of the video camera.

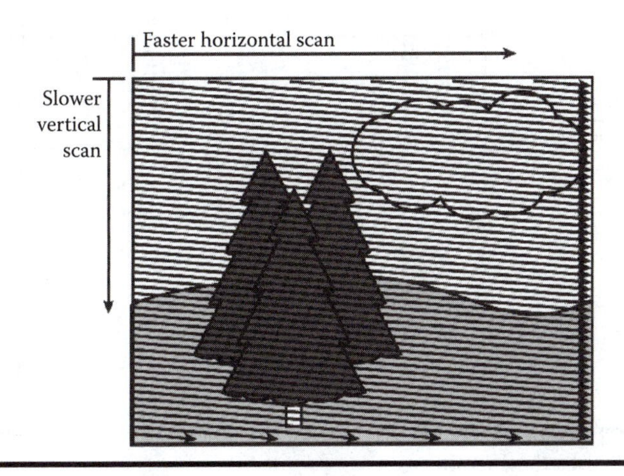

Exhibit 14.5 Electronic image scanning.

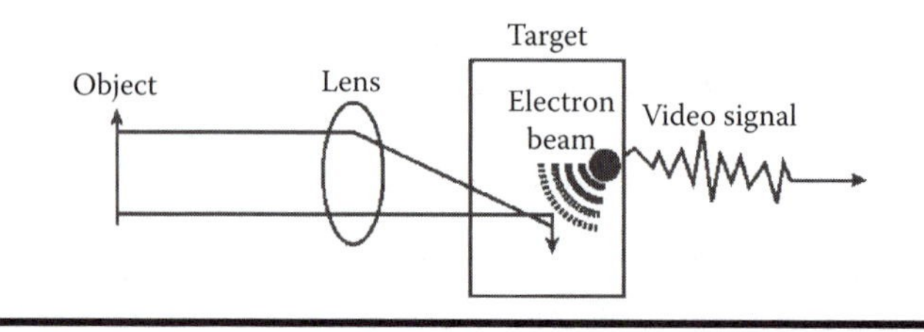

Exhibit 14.6 Scanning a focused image.

The scan line is horizontal; the scanning begins at the top left of the image and then moves to the top right. The scan line then moves vertically a short distance so that the next horizontal scan can be performed. The image or target is scanned faster in the horizontal plane and slower in the vertical plane; for video, a complete image or target is scanned every 1/30 of a second.

During a scan, the information about the light intensity of the image produces luminance (brightness) information, which is then transmitted to the receiver, where it is displayed on the receiving screen. For improved smoothness, an interlace scanning (Exhibit 14.7) method is used. Here, an image is scanned once in field 1 and a second time in field 2. This gives the appearance of an image scanned 1/60 of a second. This transfers with no visual jitter or error and provides for 525 frames, 30 frames per second, or 60 fields per second. For the vertical resolution, the image scan is capturing 367,000 dots or 525 frames (vertical) to give a clear and jitter-free image. Exhibit 14.8 shows the luminance signal of a video image.

Obviously, the image created at the camera must be synchronized perfectly with the pace of the receiver (TV monitor) for the signal to appear correctly; this

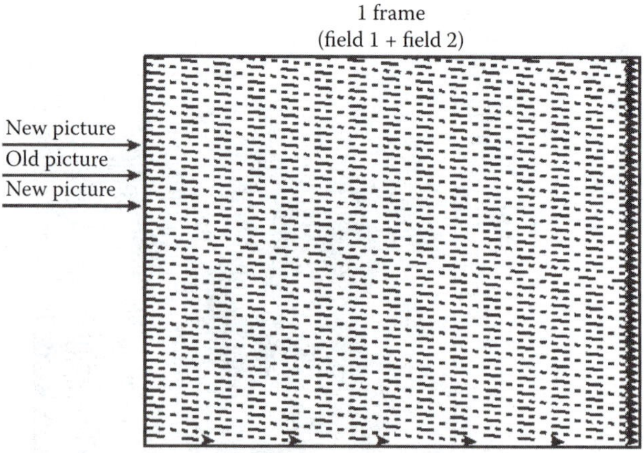

Exhibit 14.7 Interlace scanning.

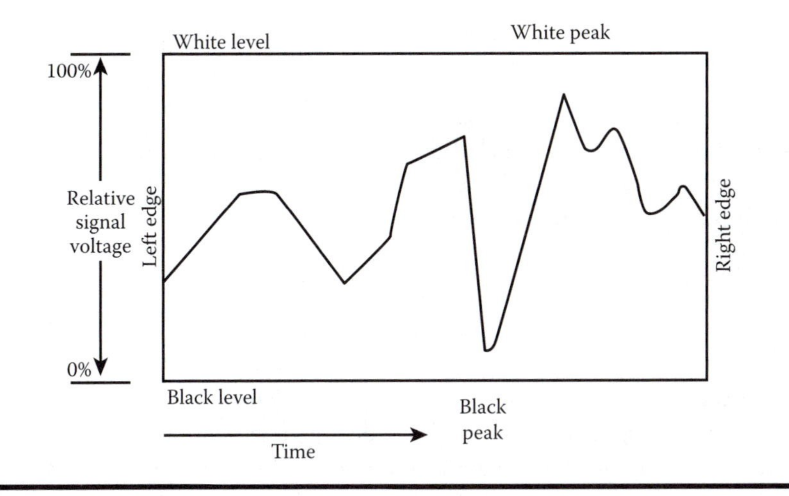

Exhibit 14.8 Luminance signal of a video image.

is achieved by horizontal and vertical synchronization pulses (Exhibit 14.9) within the video camera or transmitter and is linked with the luminance signal.

Additionally, audio (sound) information is included in the signal, along with a blanking pulse that tells the electronic beam when to turn off as it is traversing the image but not collecting or putting information on the signal. For example, in the horizontal scan, it starts at the left and gets to the right of the image, and then it must traverse back to the left-hand side to continue another horizontal sweep (at a lower vertical level).

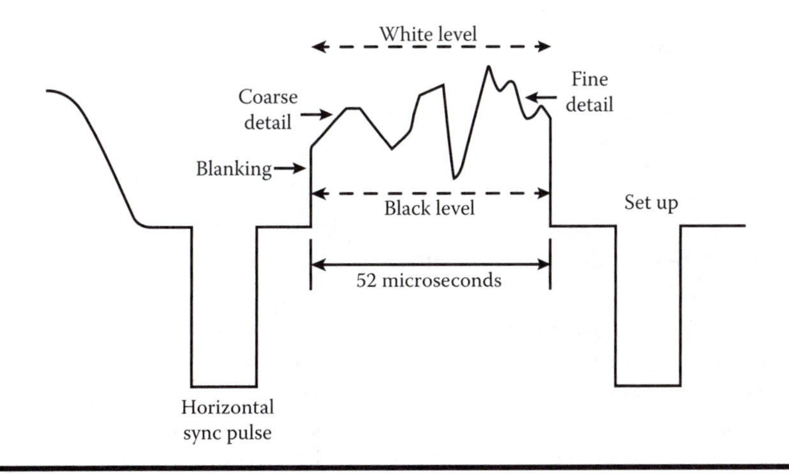

Exhibit 14.9 Luminance signal with sync pulses.

For a television signal to be complete, it must carry the following:

- Luminance information (picture brightness)
- Synchronization pulses (synchronizes images at the receiver and the transmitter)
- Blanking pulses (tells the image when to turn on and off)
- Audio and sound information

Color

So far, we have captured the luminance or brightness information of the image, but what about the color? If we just capture and display the luminance information, we have what is commonly called a black-and-white image (ranging from white to black with shades of gray in between). For most of us, the world is imbued with color, brightness, and hue. To capture these aspects of color, we must first find the primary colors, the elements that make up all color. The three primary additive colors are

- R = red
- B = blue
- G = green

Mixing the primary additive colors together produces the three primary subtractive colors:

- B + G = cyan
- G + R = yellow
- B + R = magenta

Using the additive or subtractive primary colors, all the colors (technically called *hues*) can be captured and reproduced. To capture these hues, a color video camera must be a bit different: it must separate the color into its primary elements (Exhibit 14.10).

As can be seen from Exhibit 14.10, the image is broken up into its primary additive colors by dichroic mirrors. Each of these separate colors (R, G, B) goes to a separate target camera that records the level of the specific color. As these images leave the camera, they are amplified and added together into a Y signal (composite signal), which is then transmitted and decoded at the receiving end for a color picture.

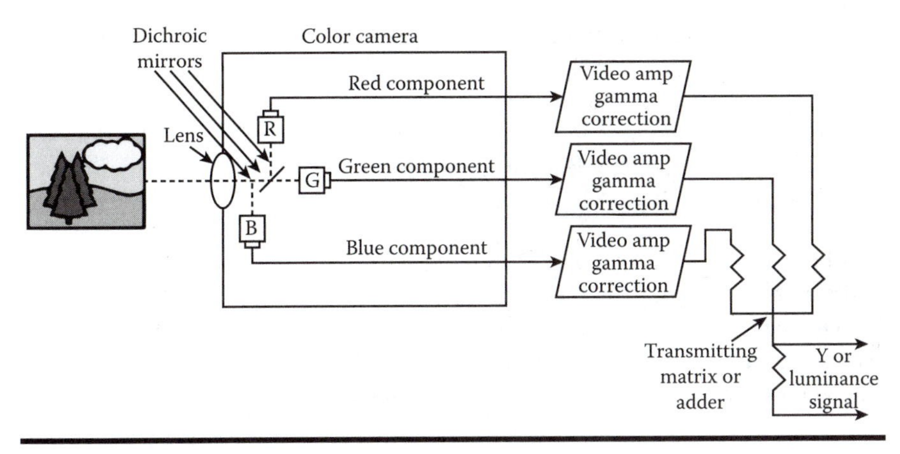

Exhibit 14.10 Block diagram of a color camera.

Transmission of Video

A black-and-white image is fairly complex (luminance, blanking pulse, synchronization pulses, and audio information); however, with color it gets even more complex (having the R, G, B information included in the signal) (Exhibit 14.11). The color images require large amounts of bandwidth to be transmitted, and each color signal requires a full 6 MHz to be transmitted properly. To make matters more confusing, TV signals are transmitted differently in various parts of the world. There are essentially three worldwide standards:

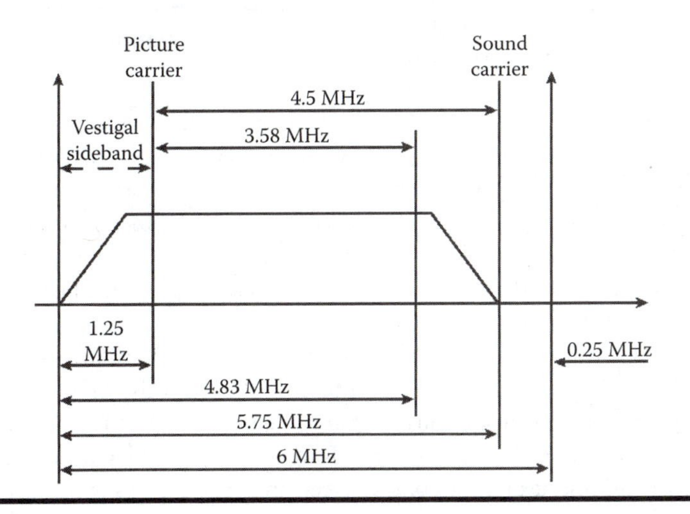

Exhibit 14.11 Color TV signal.

1. NTSC: National Television System Commission (1953)
2. PAL: Phase Alternative Line (1967)
3. SECAM: Seque Coulear Aver Memoire (1967)

These three basic standards are again modified slightly in their base form for many countries. Exhibit 14.12 shows the variations and differences among these primary standards.

Country	*Standards*
Australia	PAL-B
Austria	PAL-B
Belgium	PAL-B
Brazil	PAL-B, H
Canada	PAL-M
Chile	NTSC-M
China	NTSC-M
Colombia	PAL-D
Egypt	NTSC-M
France	SFCAM-B
Germany (East)	SECAM-L
Germany (West)	PAL-B, G
Hong Kong	PAL-B, I
Japan	NTSC-M
Korea (South)	NTSC-M
Mexico	NTSC-M
New Zealand	PAL-B
Peru	NTSC-M
Russia	SECAM-D, K
Saudi Arabia	SFCAM-B, G
Singapore	PAL-B
South Africa	PAL-I
Switzerland	PAL-B, G
Taiwan	NTSC-M
United Kingdom	PAL-I
United States	NTSC-M
Venezuela	NTSC-M

Exhibit 14.12 Worldwide TV standards.

High-Definition Television

High-definition television (HDTV) is attracting much media attention these days. Walk into any electronics store and you will see a good selection of the sets available. As we see, they are much sharper and larger (with proportions more like a movie screen than a TV screen) than the old analog TV sets. As we discussed earlier, analog TV has 525 scan lines for the image, and each image is refreshed every thirtieth of a second in an interlaced pattern. Horizontal resolution is about 500 dots for a color set. If we compare this with computer resolution, it is much lower. Computer resolution is usually 800 × 600 or 1024 × 768 pixels. We have grown to prefer the computer resolution and wonder why TV cannot do the same. HDTV basically turns your TV into a computer monitor that accepts pure digital signals and provides a high-resolution picture that is very stable and crisp. In these cases, the original analog signal must be digitized and transmitted in digital form.

The HDTV format uses a high-resolution digital television (DTV) format combined with Dolby Digital Surround Sound (AC-3). This combination creates very sharp images with theatre-quality sound (see Exhibit 14.13). The HDTV format uses a progressive scanning system. Rather than the interlaced method that shows every odd line at one scan of the screen and then follows with the even lines in another scan, progressive scanning shows the entire picture in one field every

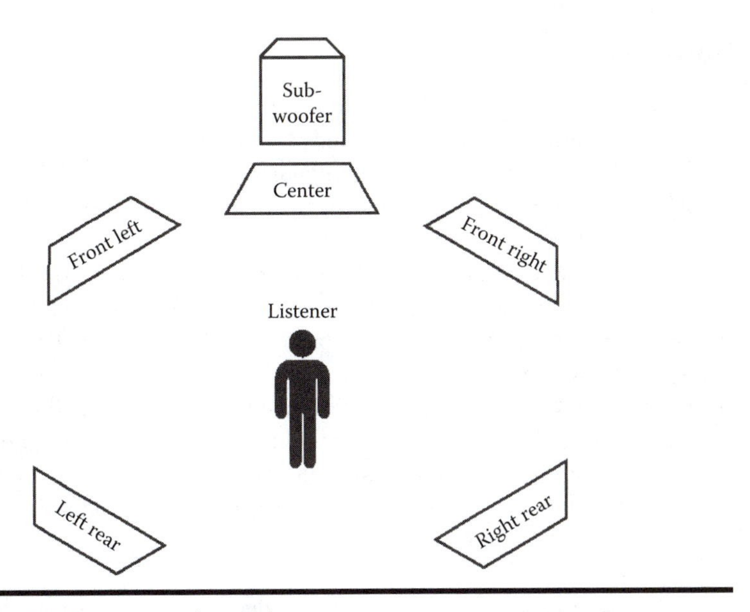

Exhibit 14.13 HDTV surround sound.

sixtieth of a second. This is a much smoother image, but it does use a bit more bandwidth.

To squeeze this additional information into a typical TV broadcast bandwidth (6 MHz), broadcasters use MPEG-2 as a compression technique. This compression software records the important parts of the image; subsequent frames record only the changes to the image and leave the rest of the image as is from the previous frame. MPEG-2 reduces the amount of data by approximately 50%. Although some detail is lost in the compression, the resulting picture is perceived to be of exceptional quality by the human eye and brain and tremendously better than traditional analog TV. See Exhibit 14.14 for an illustration of the compression process.

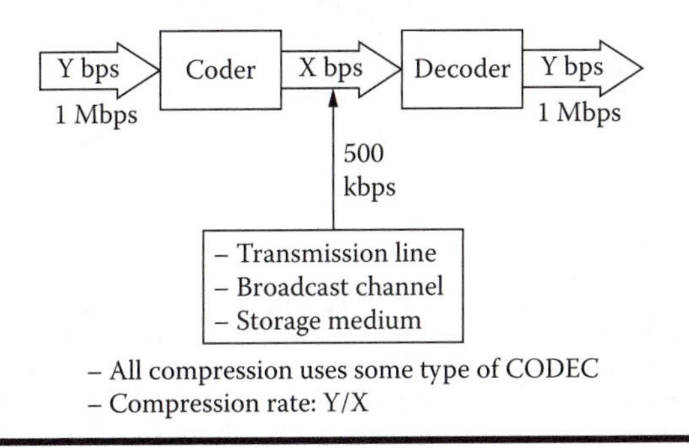

Exhibit 14.14 Digital television compression.

Summary

In this chapter, we explored the world of video. As we all know, pictures, and especially moving pictures, are an excellent way of expressing ideas and thoughts.

We looked at a common method of video communication: VC, which is common in today's business environment. Three types of VC were discussed: high, medium, and low quality.

We then explored the technical elements of video communication beginning with the psychological principle of persistence of vision. We next explored how a video image is captured, created, and synchronized. We looked at the difference between black-and-white and color images and what it takes to produce color video images. We ended this section looking at how the images are broadcast in different countries.

Exhibit 14.15 shows a case study in which the lessons of this chapter are applied. In the regulatory section, we looked at the movement in the United States toward digital television and all the elements that need to be changed for this to take effect.

Enterprise IT managers are contemplating the future of their communications networks following the recent terrorist attacks. At the same time, vendors such as Avaya continue to roll out new offerings that capitalize on user interest in IT alternatives to travel.

"After September 11, we're going to be more cautious about when to travel and when to get people on airplanes," said Chris Lauwers, CTO of Avistar Systems, a networking firm headquartered in Redwood Shores, California. "We'll just decide to travel less. And that will probably be permanent [change]."

Hence, companies such as Avistar, which links eighty employees in six cities across the United States, are continuing to look to videoconferencing to connect their workers. In addition to concerns about airline safety, many companies are considering the technology as a way to circumvent the time constraints now imposed by greater airport security.

"For short hops, it use to be that we'd just pop down from Seattle to San Francisco for a day," said Bill Hankes, a spokesman for Seattle-based service provider Internap. "I don't know that we're going to do that anymore, because now you spend two hours on each side sitting in the airport," he said.

Indeed, stories such as Robert Mason's are now far from uncommon. Mason, vice president of business development at V-SPAN, a networking company, recently held a videoconference with business partners in Austin, Texas, from V-SPAN's office in King of Prussia, Pennsylvania.

"Austin's not a convenient place to get to from Philadelphia," Mason said, "and I knew that my 6 AM flight would now be complemented by another two hours of security at each end, which would mean that it would take me twenty-four hours to do a three-hour meeting. So I did the meeting on video, and I felt pretty good when I hung up about not having take that flight back from Austin."

Exhibit 14.15 Case study: networks fill travel shoes.

Review Questions

True/False

1. Videoconferencing is simply allowing people to hear and see one another at a distance.
   a. True
   b. False
2. Because videoconferencing is half-duplex, the sender and the receiver are constantly interchanged, so at any one point in time, the person who is sending is also receiving.
   a. True
   b. False
3. High-quality videoconferencing systems are becoming very prevalent and require minimal equipment at a minimal cost.
   a. True
   b. False
4. Motion pictures use a shutter rate of 45 images per second and television images produce 60 images per second.
   a. True
   b. False
5. Electronic scanning consists of looking in vertical passes across the target image and measuring the instantaneous amplitude.
   a. True
   b. False
6. For video, a complete image or target is scanned every 1/30 of a second.
   a. True
   b. False
7. Chrominance means brightness; this information is transmitted to the receiver, where it is displayed on the receiving screen.
   a. True
   b. False
8. A complete television signal carries luminance information, synchronization pulses, blanking pulses, and audio information.
   a. True
   b. False
9. The three primary additive colors are red, yellow, and green.
   a. True
   b. False
10. Each color signal requires a full 6 MHz to be transmitted properly.
    a. True
    b. False

Multiple Choice

1. The standard currently used in the United States to transmit video signals is
   a. PAL
   b. SECAM
   c. NTSC
   d. None of the above
2. The technology used to squeeze a lot of information into a small space by using mathematical algorithms to reduce bandwidth is
   a. Transmission
   b. Compression
   c. Progression
   d. Synchronization
3. What tells the electronic beam when to turn off as it is traversing the image but not collecting or putting information on?
   a. Blanking pulses
   b. Y signal
   c. Synchronization pulses
   d. Luminance signal
4. The correct interlace scanning method provides for
   a. 400 frames, 30 frames per second, or 60 fields per second
   b. 525 frames, 30 frames per second, or 30 fields per second
   c. 400 frames, 60 frames per second, or 60 fields per second
   d. 525 frames, 30 frames per second, or 60 fields per second
5. For video, a complete image or target is scanned
   a. Every 1/30 of a second
   b. Every 1/60 of a second
   c. Every 1/100 of a second
   d. Every 1/50 of a second
6. The rapid presentation of frames of video information to give the illusion of smooth motion is
   a. Continuity of action
   b. Interlace scanning
   c. Persistence of vision
   d. Electronic scanning
7. The videoconferencing systems that are becoming very prevalent and require minimal equipment at a minimal cost are
   a. Low quality
   b. Medium quality
   c. High quality
   d. None of the above

8. The transmission media typically used in medium-quality videoconferencing is
   a. Internet
   b. ISDN
   c. Fiber optics
   d. Satellite
9. The image of a video signal is broken up into its primary additive colors by
   a. Reflective mirrors
   b. Image mirrors
   c. Scanned mirrors
   d. Dichroic mirrors

Short Essay Questions

1. What are two distinct advantages of the digital video format?
2. Explain the concept of compression used in digital television.
3. What are the three basic elements of videoconferencing?
4. What is the main difference between film pictures and video pictures?
5. What are the four parts that make up a complete television signal?
6. What are the three primary additive colors? Primary subtractive colors?
7. Compare the three essential qualities of videoconferencing systems.
8. Explain the principle of persistence of vision.
9. Describe the process of electronic image scanning.

Chapter 15

Digital Media

Introduction

This chapter will take a different path from the previous chapters. Rather than concentrating on the medium, as did other chapters, this chapter will focus more on the message. The message is digital, or more commonly known as digital media.

This chapter will look at a few of the more popular digital media sources available. Specifically, we will look at

- Digital photography
- Digital video
- Digital audio
- Digital gaming

We will look at each of these digital media forms and cover them by asking the following questions:

- Why did this digital media form come to be? What are the advantages and disadvantages over the analog version?
- What are the technical differences between the digital form and the analog form? (This assumes that you know the old analog version.)
- What new opportunities does this digital media offer?

Digital Media

Florida's digital media industry association defines digital media effectively: "the creative convergence of digital arts, science, technology and the business of human expression, communication, social institute and education."

Some key elements of the definition are described in the following subsections.

Convergence

Digital media has come together as a result of convergence between many different fields and subjects. What used to be walls between disciplines have now broken down, resulting in a new core for digital media. What were once disparate disciplines of science, technology, and the arts now are the tools and palette of the new artists and communicators. However, with this new digital media come a new business perspective and a different business angle for the typical media business: TV, radio, newspapers, etc. Corporate America sees many new services and revenue opportunities in their digital media future. Digital media has also created whole new industries, electronic gaming to name one, which has sprouted from a seed into a giant tree, apparently overnight, as a result of this convergence.

1. *Digital*—Digital media implies that the product, at some point in its life, was in an all-digital form. Although digital theories have been around since the Babylonians, with work done by Carl Gauss and John Napier in the early 1800s, it was the work of Francis Bacon (1620) with the first binary alphabet that was the turning point for digital logic. However, digital media had to wait until the electronic age before it could flourish. These new types of digital media imply an electronic shape that operates on digital codes.
2. *Media*—Digital media usually implies a baseline format. Formats are used for digital information and, in the current world, come in various shapes, sizes, and colors in each of the disciplines of digital media. These disciplines are digital audio, digital video, and other digital content that can be created, adapted, and distributed via digital information processing equipment.

The impact digital media represents is either profound and revolutionary or just another ho-hum medium for us to communicate, express ourselves, and educate people. The viewpoint depends on your perspective, age, history, and background. For most people over the age of 50, digital media is just a new palette, a new color, or a new instrument in the creation of paintings, pictures, sound, or visuals. To those under 30, digital media is revolutionary. It brings together the arts and the sciences, which should never have been separated and should be naturally integrated. Additionally, those under 30 see the term *digital media* as redundant. All media is digital or should be. This is the digital media perspective, where anything is possible.

Exhibit 15.1 Artwork by Roger Langridge.

Along with the new field of digital media come new artists, inventors, and pioneers. Some of the new pioneers in this field, the people who have made it work, are

- Andrew Wildman
- Raymond Mullikin
- Roger Langridge (Exhibit 15.1)
- Ben Hatke
- Mathew Forsythe
- Rob Feldman
- Scott Dulton
- Andrew Dalab
- Ernie Colon

These people are artists and scientists, engineers who are not part of the digital revolution but are the digital revolution and are asking questions and creating messages, meanings, and products and services that were earlier unthinkable.

The world of digital media is not just for the appreciation of the young and the pioneers. It has also influenced our universities and colleges, the typical bastions of tradition, as well. A 50-state survey funded by the Lilly Foundation and implemented by the International Digital Media and Arts Association (http://www.idmaa.org) investigated how broad-based the digital media and arts phenomena had become. Some notable data from this survey include the following:

- At least 10% of all postsecondary institutes in the United States offer digital media courses; such courses are part of almost 400 collegiate programs.
- Most digital media programs are interdisciplinary or multidisciplinary programs. These are typically hard to create in a traditional academic setting.
- There are currently 30,000 students attending digital media coursework throughout the United States.
- Digital media courses come from a very wide range of academic areas, including gaming, computer science, television, wireless, graphics design, technology, theater, architecture, and photography; the list goes on.

So what does this mean to us? The digital revolution, with its child, digital media, is widespread and growing and has support from the young, the old, and corporate America. All see different benefits and representations of digital media, but all see it as the future.

Digital Photography

Advantages of the Digital Form over Analog Form

Digital photography arose quickly and effectively to literally obliterate the older analog camera and film combination (hereafter called *wet photography*). Even the mighty Kodak Corporation, the founding fathers of wet photography, was taken by surprise. But why? What did digital photography (*dry photography*) offer to the masses that allowed, promoted, and created such a drastic and quick change? Two reasons are considered in the following subsections.

Convenience

With the older wet photography technology, one had to buy film, load it in the camera, take the picture, rewind the film, bring the film in for development, wait, pick the prints up from the store, buy a photo album, and store the prints. Then, upon looking for the picture a year later, one had to look in various albums, in various places, for the picture wanted. More important, the negative, necessary for a copy, was nowhere to be found. Dry photography solved most of these issues while also providing instant feedback so that one could see if Uncle Fred had his eyes closed in the picture that was just taken.

Personalization

In 2006, *Time* magazine announced its man of the year: it was you, and me, and everybody else. The ME generation, mass customization, has created a world in which we can all see ourselves as individuals in the center of the universe and adapt our surroundings in the way we like it. Digital photography offers this personalization feature so that we can change the prevalent reality. Once in the form of 1's and 0's, a picture could easily be edited so that Uncle Fred is on the left rather than on the right of the picture.

Yes, wet photography offered many darkroom techniques for image manipulation, but these tricks were left to the professionals, the darkroom experts who would work their magic. With digital photography, an electronic image can be taken and, with the help of any of the plethora of editing tools listed in Exhibit 15.2, changed in any way.

Technical Differences between Digital and Analog Forms

There are many more similarities between wet and dry photography than one can imagine (lens, optics, lighting technique, poses, composition, etc.), but here we will concentrate on the differences. What makes digital photography different? Certainly, it would behoove the reader to read up on general photography before delving deeply into digital photography. In the following subsections, we list some of the key technological points that differentiate dry from wet photography.

Name	*Cost*	*Pro*	*Cons*
Professional			
Adobe Photoshop	$649.00	Supports Files: RAW Imports BMP GIF JPEG JPEG 2000 PNG TIFF PSD Supports: Windows Mac OS X Sized printing, sharpening, color correction, lens correction is possible	Does not support: Linux BSD Unix
Micro media Fireworks	$299.00	Supports Windows Mac OS X Histogram, Scripting, selection editing is possible	Does not support: Linux BSD Unix Lens correction and color correction is not possible
Deneba Canvas	$349.99	It supports both raster graphics and vector graphics. Histogram, Scripting, selection editing is possible	Does not support: Linux BSD Unix. Sized printing, sharpening, color correction, lens correction is not possible
Alias Sketch Book Pro	$179.00	Supports Windows Mac OS X Layers and selection editing is possible	Does not support: Linux, BSD, Unix Has no features such: Histogram Scripting HDR Does not support Grayscale and CMYK
Photogenics	$699.00	Supports: Windows and Linux. Layers and selection editing and histogram is possible	Does not support Mac OS X and Unix. Lens correction and color correction is not possible
Advanced			
Ability Photopaint	$29.99	Supports Files: BMP GIF JPEG PNG TIFF PSD	Does not support JPEG 2000 file
Corel Paint Shop Pro	$99.99	Supports Files: sRGB Adobe RGB Indexed Grayscale CMYK	Does not support: Mac OS X, Linux, BSD, Unix
Corel PHOTO-PAINT Helicon Filter	$299.00 $105	Histogram, Scripting HDR, Lens correction and color correction is possible Supports HDR Retouching Resizing Noise removal Lens correction Sized Printing Sharpening Color correction Plug-in support	Does not support: Mac OS X, Linux, BSD, Unix Does not support Linux, BSD, and Unix. Selection editing, Layers and scripting is not possible

Exhibit 15.2 Common photography editing tools: simple, advanced, and professional.

(Continued)

GIMP	Free	Supports most Color spaces Supports Files: sRGB Adobe RGB Indexed Grayscale CMYK	Does not support Image Library Does not support JPEG 2000 file
Simple			
Adobe Photoshop Album starter Editor	Free	Supports Windows	Does not support: Mac OS X, Linux, BSD, Unix
Picasa	Free	Sized printing, sharpening, color correction, image library is possible	Lens correction, noise removal, scripting, retouching and layers is not possible
Pixia	Free	Selection editing, layers and histogram is possible	Sized printing, sharpening, color correction, image library is not possible
iPhoto	Free	Sized printing, sharpening, color correction, image library is possible	Selection editing and layers is not possible
Paint.NET	Free	Selection editing, layers, retouching and resizing and sharpening is possible	Noise removal, lens correction and sized printing, histogram and scripting is not possible

Exhibit 15.2 (Continued) Common photography editing tools: simple, advanced, and professional.

Camera Sensor

Rather than using one-shot photographic film, a digital camera has a sensor to digitally record the image. The sensors (Exhibit 15.3) are typically complementary metal oxide semiconductor charged-coupled devices (CCDs), arranged with color filters (Exhibit 15.4). On the surface (the grid), each light sensor position is called a pixel (picture element). For a typical 35-mm film size, it takes 1.5 to 3 million pixels for comparable quality. Each pixel *sees* the light hitting it and reads it in a digitized bit form of 8, 10, or 12 bits (bit resolution). Color filters are used

Exhibit 15.3 Photographic sensor.

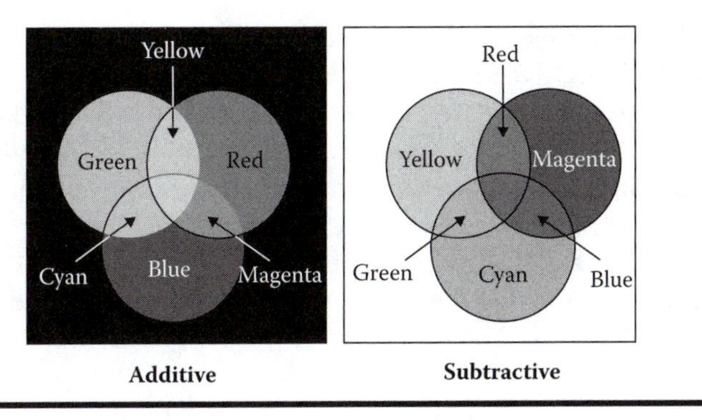

Exhibit 15.4 Sensor filters.

to capture the different sensitivities of either the additive or subtractive primary colors. If the camera is additive, the additive primary filters red, green, and blue are used. RGB added together makes white light. For the subtractive concept, it is cyan, magenta, and yellow filters that are used. Cyan, magenta, and yellow together make black.

Wet photography film has its issues (reciprocity failure is one) and so does its counterpart in dry photography, the sensor. The camera sensor develops noise. Noise is random and pseudorandom bits of electronic energy picked up by the pixels and amplified to become part of the digital image. Sensor noise is increased by long exposure times, overexposure, and high temperatures.

Storage Media

Once the picture has been taken in the camera, it must be stored, not on film but in a binary holder. The typical binary storage areas for digital cameras are compact flash (CF) and secure digital (SD) cards. A CF card is a memory device that retains data without direct power, as well as defines the physical format and interface of the card. Inside the card are solid-state memory chips and a chip controller. CF cards come in two different sizes, Type 1 and Type 2. The only difference is in their thickness, which usually determines their storage capacity.

The most important performance characteristic of a CF card is the speed at which it can read and write data. Usually, the write speed is the more important factor because it determines the speed at which a digital camera can shoot a picture.

Speed is often given as a multiplier (such as 4×, 24×, 40×, 80×, and 133×). 1× is usually defined as a write speed of 150 kbps. 12× is about the slowest you can buy, and 133× is the fastest. Read and write speeds are usually similar.

There are two different types of memory cells used in the CF memory. The fastest is called a single-level cell (SLC), which stores 1 bit in each cell. The slower (but cheaper) architecture is called a multilevel cell (MLC), which stores two bits in each cell. SLC technology also uses a little less power than MLC, so that is another advantage. The main advantage of an MLC is its lower cost.

These memory cards come in a variety of sizes, from 4 MB to multiple gigabytes of data storage. Depending on the quality of the camera and the photograph you take, a single image can range from 50 kB to a few megabytes. The card dictates how many pictures your camera can take before you must replace it.

Once stored in a digital form on the camera, the picture can be later moved to any of the more common digital formats such as any smart medium, memory sticks, your hard drive, or a compact disc.

Compression

Digital images tend to take up a large amount of storage space. To reduce this space requirement, compression is used. Various compression algorithms are used with digital photography. The more common tools are as follows:

JPEG: Joint Photographic Experts Group (JPEG) is a compression algorithm that compresses without a significant loss of picture quality. JPEG is an algorithm designed to work with continuous tone photographic images; it takes image data and compresses them in a lossy manner. Lossy means some information is lost, and therefore, picture quality is reduced. The more you compress, the smaller the file becomes but at the expense of information (some information is lost) and, hence, picture quality. In practice, you can reduce file size by a factor of 10 and still get a very high-quality image, about as good as the uncompressed image. With JPEG you can reduce the file size by a factor of 40 or more, but the image quality suffers.

With 10:1 compression, an 8-bit file would be about 900 kB in size rather than 9 MB without using JPEG. This is a big savings in storage size, and smaller files are faster to send from the digital camera to a computer.

TIFF: Tagged image file format (TIFF) is a lossless way of saving files. TIFF retains all the original information, but at the expense of file size; the files are bigger. TIFF files can also be used to save 16-bit data; JPEG files can only save 8-bit data.

Proprietary Compression

Many of the popular cameras offer a proprietary way to save the actual data generated by the sensor. Canon calls their version of this *RAW*; Nikon calls it *NEF*. These files are compressed in a *nonlossy* manner (Exhibit 15.5). They are smaller than TIFF files but larger than JPEGs. Typically, they achieve a compression of around 6:1. The disadvantage of these formats is that the image must be converted to either JPEG or TIFF for most software to be able to display and manipulate them.

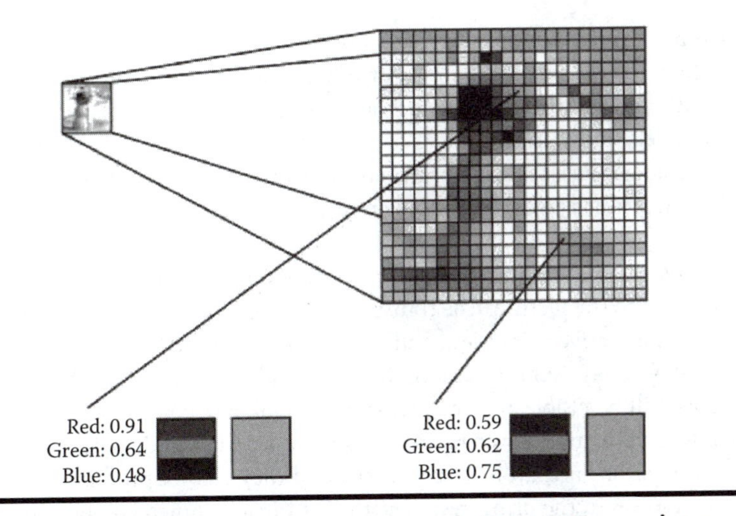

Exhibit 15.5 Comparison pictures: uncompressed, JPEG, TIFF, proprietary.

Digital Darkroom Techniques

Similar to its counterpart in wet photography, digital darkroom techniques can correct for common flaws in pictures. However, with dry photography, these corrections can take place in the camera or with any simple editing tool. The typical picture flaws that amateurs have to deal with (by their own creation) are

- Fixing under- or overexposed pictures
- Adapting color, contrast, dodging, or burning
- Fixing red eye
- Cropping out unwanted elements

Printing

There are many printing technologies available to turn the digital picture into something we usually recognize as a photograph (a wallet photo, a 3 × 5 picture, or an 8 × 10 photo). With picture printing, we must understand ppi and dpi.

Pixels per inch (ppi) is a term exclusively used for picture printing. If you take an image that is 800 pixels wide and 600 pixels high, and you print it with a setting of 100 ppi, the print will be 8 inches wide by 6 inches high. If you print at 200 ppi, you get a print 4 inches wide by 3 inches high. The print at 200 ppi will be smaller but higher in quality. The common format of 320 ppi is thought to be the best. Above 320 ppi, it is hard to see any improvement in image quality. Below 180 ppi, there is a noticeable quality drop.

Dots per inch (dpi) is a property of a printer, not of a digital image. It is a measure of how finely spaced the drops of ink are in a print. Printer settings of 360, 720, 1440, and 2880 dpi are common. The differences between these are often unnoticeable. Most consumers are happy with 360 dpi. Changing the dpi setting does *not* change the size of the print; ppi controls that. Dpi controls print quality.

There are three types of technologies used for printing:

Dye sublimation—With dye sublimation, transparent ribbons are heated and bonded to the picture. The transparent ribbons are yellow, magenta, and cyan (the subtractive primaries), and together they can make any color of the rainbow. With dye sublimation, the colors will not be particularly dense, but no dots will be visible and the pictures will look like conventional photographs. Dye sublimation printers are the simplest to use and maintain. They cost more per print than ink-jet printers, and they cannot produce the best quality. However, the prints resemble those from a commercial photofinisher, and they tend to be durable and long-lived.

Ink jet—The highest print quality possible comes from inks. Ink-jet printers use the same three subtractive colors as dye sublimation, but the inks are more opaque. The colors are applied adjacent to one another. Often, black is also used as an ink color to improve contrast and harden the edges. Standard consumer-quality ink-jet photo printers use four inks: yellow, magenta, cyan, and black. These four are sufficient to form a complete range of colors. If you want better, the best-quality photo printers add light magenta and light cyan to the other four colors.

Laser—Laser printers are very fast and very cheap per print, but quality suffers.

Advanced Darkroom Technique

Beyond the fixing of common flaws in pictures, as covered earlier, digital photography opens up new realms in adapting pictures to what we want to see rather than what we actually saw. For example,

Remove and modify background—Rather than the family standing on the front lawn, why not substitute the lawn with Paris, Istanbul, or the moon. A relatively easy merging of pictures could achieve this effect.

Correction of lens perspective distortion—As objects recede in the distance, they come to a point of perspective. Mentally, when we see this, our mind's eye adapts it the way it should be. However, the camera does not lie. It captures the distortion. This can be easily corrected (Exhibit 15.6a and b).

Various speed effects—Our mind's eye sees things moving in a continuum and does not distort them as they are moving. A still camera captures but a moment in time, and often during that moment, the object is moving and therefore blurry. This also could be corrected.

Exhibit 15.6 (a) Uncorrected lens perspective. (b) Corrected lens perspective.

New Opportunities

With the advent of any new policy, law, or technology, or any major change, new opportunities present themselves. Here are some of the new services, products, and ideas that have arisen as a direct result of digital photography that would not be possible with wet photography:

Picture sharing—Shutterfly and many other companies can send your pictures to anybody who cares to look at them on the web. This sharing application has really taken off and gives people the option of boring thousands with their vacation pictures.

Image archiving—Traditional pictures, once on paper, tend to degrade over time. This is especially true of color pictures. Even the wet photography negatives are prone to loss and degradation and, because of their one-shot nature, cannot be replicated. Digital photography offers archival permanence and creation of many images (without loss of quality), ease of finding (by sorting the name), and can be put in a consumer-friendly format for easy showing (videotapes, CDs, and DVDs).

Visual note taking—This is an application that is increasing in popularity, but corporate America has yet to figure out a way to make money off it (beyond the camera sales). For example, you go to Disney World, and after you park, you take a picture of where your car is (I always forget). During a conference, you take pictures of the presenters' slides or notes on the whiteboard. You take pictures of faces at a family reunion, so you could remember everybody the next year. The possibilities are endless, and with the compact size and low cost of cameras, this is a homegrown use that is certainly growing.

Digital Video

Digital Video: Genesis and Advantages over Analog Video

Similar to still photography, digital video photography has risen up very fast. Consumers, and even professionals, are adapting to this new era with enthusiasm. Sales of camcorders have exploded. Why did this trend begin? What does it offer the consumer?

Cost—Digital camcorders cost from 5 to 10 times less than their comparable analog film cameras. This promotes sales and use. Additionally, there are no additional costs due to film processing, a film projector, and a dark room.

Flexibility—A digital camcorder easily changes from a camera to a playback device, thus offering the creator the option of instant feedback. Additionally, this camera tends to be smaller and lighter than its analog counterpart and therefore can be carried anywhere.

Video editing—This is greatly simplified; scissors, tape, and specialized equipment are not needed. The pictures can be easily transferred to a computer, put into a digital editor, and completely redone, remixed, and edited (with no loss of image quality as in the analog world).

Technical Differences between the Digital and Analog Forms

Plate—The movie camera's film is a CCD plate. The plate captures the light falling on it and converts it to an electronic level representing the amount of light. Similar to a still camera, a CCD is divided into a grid with small pixel elements. Most camcorders usually have a 1.3 or 1.4 in. CCD having between 270,000 and 880,000 pixels. This is far less than a still camera, which has 3 million pixels.

Two key criteria of the CCD plate are the resolution and the light sensitivity necessary to shoot a picture. Light sensitivity is called the lux rating. Lux ratings range from 1 to 5 with the lower number being better.

Another criterion of the plate is the signal-to-noise ratio or the SN ratio. The signals coming from the CCD are sampled and then amplified. In the amplification comes noise (unwanted signals). A good amplifier/CCD combination has a high SN ratio, allowing the noise and picture to be easily separated. Cheaper cameras have a low SN ratio, which allows noise to enter the picture easily.

Video storage—Instead of rolls or cans of film, digital video is usually stored in a Mini DV tape, although there are other formats. Review Exhibit 15.7 for comparison.

Connections—Although you can view all of your digital videos on the camera itself, usually the video is watched on a TV monitor or PC. There are many ways to implement this transfer of information. The major ones follow:

IEEE 1394 Port (called FireWire by Apple and i-Link by Sony)—This is a cable that transfers video from the camcorder to the computer or between matching camcorders (Exhibit 15.8).

Certainly, FireWire is the most popular method of image transfer as it has many attributes necessary for this field. Below are some examples:

Speed—It is capable of 400 Mbps throughput and continues to speed up with new enhancements. It is capable of two simultaneous channels of broadcast-quality, full motion (30 frames per second), and CD-quality stereo audio.

Fast available connection—It is part of the *plug-n-play* suite, so all one needs to do is plug in the wire and the rest is done for you. Additionally, you do not have to power down equipment when you plug it in. It is hot pluggable.

Thin cable—It is a light, thin cord that makes it easy to pack; it is also very ergonomically balanced.

Daisy chaining—You can daisy-chain 63 devices using one physical port.

Distribute power—The cord is capable of distributing a low level of power for low-consumption devices.

Infrared—Some camcorders also use a wireless infrared connection to send the video from the camcorder directly to a television set.

Format	*Capacity*	*Size*	*Manufacturer*	*Cost/MB*
Mini DV	Single Layer: 4.7 GB	12 cm Single sided	Warner and Cyber Home	$ 1.5/GB
	Double Layer: 9.4 GB	12 cm Double sided		
HD DVD	Single Layer: 15 GB	12 cm Single sided	Toshiba	$ 1.97/GG
	Single Layer: 30 GB	12 cm Double sided		$ 1.18/GB
Blu-ray Disc	Single Layer: 27 GB	12 cm Single sided	Sony and Philips	$ 1.31/GB
	Single Layer: 7.8 GB	8 cm Single sided		
DVD+R	Single Layer: 4.7 GB/8.5 GB	12 cm Single sided	Sony TDK Philips Mitsubishi Memorex	$ .05/GB
DVD-RAM	Single Layer: 4.7 GB	12/8 cm Single sided	Panasonic	$ 1.91/GB
	Double Layer: 9.4 GB			

Exhibit 15.7 Video storage formats.

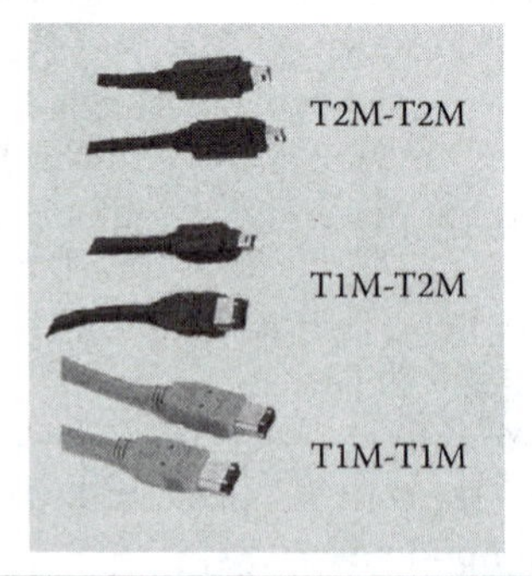

Exhibit 15.8 FireWire ends (6–6, 4–4, 4–6).

S-Video—The S-Video output is used to play or transfer images from a camcorder to a monitor, television, or an analog VHS tape machine. S-Video is an analog transfer of information, but in one of the better forms for analog transfer. The S-Video signal carries information for color (Chroma) and brightness (Luminescence) on separate lines within the cable.

New Opportunities

With these new camcorders, digital video offers a host of new opportunities. Some of these are as follows:

Instant shooting and editing—Now you can shoot and do limited editing in the camera, so the product is a semiprofessional-looking video available within minutes after the event really happened.

Document everything—With the small size and price, the camcorder offers people the ability to document everything. Is this necessary? Maybe not, but we have all seen or heard the effects of being on camera, of seeing someone on camera that is doing something wrong (i.e., the Rodney King film).

Note taking—This is similar to still photography but offers more continuity. You go see a lecture from a famous Nobel laureate, and you want to capture his/her words and images forever. The camcorder offers this possibility.

Digital Audio

Digital audio has not been as big a hit with the consumer as digital photography. Whereas just about everybody, young and old, is replacing their *wet* camera with a digital one, most people are not replacing their audio system with a digital audio system. However, digital audio has established a foothold in the consumer market and has hit big time in the personal device arena.

Advantages over Analog Version

What is digital audio? It is a digital signal that represents analog music and sound. Although we speak, most instruments play, and we hear in analog, digital audio has made an entrance in the consumer market. Why has this happened? For a number of reasons:

1. *Recording music*—Although most instruments are analog in nature and we still use air to get the sound to a microphone, digital recording is done as it tends to be noise free and can be easily manipulated.
2. *Manipulation*—Once recorded in digital form, the sound is easier to manipulate, enhance, and change than in analog form.

3. *Mass production*—A digital recording loses nothing as it changes hands from generation to generation (unlike analog recording). Therefore, the original music is preserved. Additionally, the mass production of digital audio, in the form of compact discs, streaming audio, or MP3 files, is cheaper and easier.
4. *Distribution*—Once in digital form, it is easier to mass-produce CDs and ship files around on the Internet. This leads to reduced costs, and loss and noise do not have to be factored in at every stage, unlike analog.
5. *Quality*—The debate regarding quality is still on. Although there are numerous objective reports that show that either digital music or analog music is better, the ultimate decision is subjective. There are many different flavors of digital recording for us to fight over, such as
   a. *DDD*—Digitally recorded, mixed, and distributed; the ultimate digital audio recording.
   b. *ADD*—Recorded in analog, mixed digitally, and digitally distributed. This is often done on old recordings that cannot be rerecorded.

Technical Differences between Digital and Analog Forms

The first step in digital audio is its conversion into digital form. Although not necessary for every instrument (such as a software synthesizer), this step is necessary for all voice and 99% of instruments. Once received by the microphone in analog form, the signal is sent through an analog-to-digital (A-D) conversion. This conversion is very similar to the A-D conversion talked about in Chapter 3, so we will not reconsider it here. It does involve a sampling rate and a bit resolution. For most CDs, the sampling rate is 44.1 kHz, and the bit resolution is 16 bits.

After this step, the digital signal is processed. Here the music is filtered, effects applied, and the music stored on a CD, iPod file, hard drive, or USB device.

As we learned earlier, digital signals take up more space than their analog counterpart. This is also true of digital music, and therefore, data compression is often used. Typically, MP3 is used as the compression algorithm for audio.

The final step is the digital-to-analog (D-A) conversion. Here the digital signal is changed back into analog form so that it can drive a loudspeaker or a set of headphones. There are many tricks in this stage to make the music sound better or make up for inherent flaws of digital music. These include the following:

1. *Oversampling*—With digital signal processing, the normal sampling rate is twice the highest frequency. So, for a signal with the highest frequency of 20 kHz, the sampling frequency would be 40 kHz. Oversampling uses a sampling rate that is substantially higher than this. This results in a higher quality of music.
2. *Upsampling*—This is the process of taking a digital signal and the original sampling rate and subjecting it to a higher sampling rate. This results in a higher bit rate and better music quality.

3. *Downsampling*—Downsampling, sometimes called subsampling, is the opposite of upsampling. It involves taking the original digital signal with its original sampling rate and subjecting it to a lower sampling rate. This results in a reduced bit rate and therefore lowers the signal quality; however, the signal becomes easier to transport.

Digital audio is associated with new interfaces:

1. *AC97*—AC97 stands for Audio Codec 97. The standard was developed by Intel in 1997 and is used mainly in modems and sound cards. The two main components are the AC97 digital controller and the AC97 audio and modem codec. AC97 supports 96,000 samples/s in 20-bit stereo resolution and 48,000 samples/s in 20-bit stereo.
2. *MIDI*—Short for Musical Instrument Digital Interface, it is a standard protocol for the exchange of musical information. The exchange takes place between instruments (guitars, etc.), synthesizers, and PCs. MIDI has been around for a long time, with its first appearance in the early 1980s.

New Opportunities

The introduction of any new technology brings new opportunities. Also, digital audio is so new that we cannot even ask the right questions to cover the unlimited possibilities. As of now, the new opportunities are as follows:

1. *Music sharing*—Music sharing started off with a bang and then got caught in the legal net of the music companies. In the form of peer-to-peer networks, such as Napster and the newer versions, the networks allowed, promoted, and encouraged those who had music to share to do so with others who wanted it. Once in digital form, music is easy to transport via the Internet.
2. *Digital audio players*—Need we say more? Starting with portable CD players and progressing to MP3 players and the famous iPod, digital audio has totally opened up this field.
3. *Digital audio workstations*—Don't like the drum beat for "Knocking on Heavens Door?" Thought you could sing "She Loves You" better than the Beatles? Digital audio workstations allow manipulation of digital music. These range from freeware versions all the way up to professional software and equipment that cost tens of thousands of dollars.
4. *Compact disc*—Once, audio was limited to the music or living room of the house. With the compact disc came portability and flexibility.
5. *Satellite radio*—Satellite radio has become a hit. It merges digital audio and satellite technology. Once you get used to commercial-free satellite radio, it is hard to return to over-the-air analog. Satellite radio has made a significant penetration into the consumer market.

Most of us listen to the radio as we drive to work or for pleasure. Most of us also preset the radio buttons to fit our moods during the day. On a long trip, this becomes a problem. Radio stations typically carry for about 30 to 40 miles, and on a long trip we constantly change the radio frequency to pick up *local stations* as we travel the interstate road. With satellite radio, the radio station transmitter is located in orbit, 22,000 miles above the earth's surface. Because of this location, it can cover the whole of the United States. This means you can listen to the same station all the way from Los Angeles to New York.

It all started in 1992 with the Federal Communications Commission (FCC) reserving part of the 2.3 GHz frequency band for Digital Audio Radio Service (DARS). In 1997, two companies were granted licenses to broadcast. These companies are now XM and Sirus Radio.

In the United States, XM and Sirus Satellite Radio are the two big players offering satellite radio service now. In the world market, WorldSpace is the global leader. Satellite radio is different from analog radio in two major ways: (1) programming and (2) technology.

At the programming end, satellite radio providers differentiate their service from regular analog radio by a few factors: (1) near-CD-quality sound, so the audio quality is superior to that of regular FM; (2) limited or no commercials; (3) genre stations (e.g., a station devoted to 1960s music); and (4) a subscription fee (usually $13 a month).

From a technology standpoint, satellite radio comprises three major components. The first component is the head end. These are the earth-based studios that assemble the music, encode it, and beam it up to the satellites. The second component of the service is the satellites, which take the communication from the head end and retransmit it down to all the parties. XM uses two satellites, called *rock* and *roll*, placed in geosynchronous orbit to cover the North American continent. Sirus uses three satellites in an elliptical satellite constellation. The third unique technology component is the receiver. The receiver is made up of two components. An antenna, similar to a car phone antenna, receives the 2.3-GHz signal, amplifies it, and passes it on to the chipset. The chipset decodes the transmission down to a more normal frequency and performs any necessary decoding. Both companies make these receivers for home and automotive use and are licensing their receivers to other manufacturers.

6. *Internet radio*—Why Internet radio? Why use an expensive PC as a radio? There are numerous reasons:
   a. One can listen to the channel offering music one grew up with or are accustomed to.
   b. A geographically diverse constituent can be united via private broadcast.
   c. A radio listener hears an ad for a computer printer and places an order immediately using the same medium on which he/she heard the ad.

d. Not limited to just audio (images and alike).
e. Unlimited range for broadcasts.
f. No need to buy or use limited FCC spectrum.

Internet radio has been around since the late 1990s. Traditional radio broadcasters have used the Internet to simulcast their programming. However, Internet radio is undergoing a revolution that will broaden its access from the desktop computer to access anywhere, anytime, and expand its programming from traditional broadcasters to individuals, organizations, and government.

Radio broadcasting began in the early 1920s, but it was not until the introduction of the transistor radio in 1954 that radio became available in mobile situations. Internet radio is in much the same place. Until the twenty-first century, the only way to obtain radio broadcasts over the Internet was through your PC. That will soon change, as wireless connectivity will feed Internet broadcasts to car radios, cell phones, and all personal digital assistants (PDAs).

Internet radio programming offers a *wide spectrum of broadcast genres*, particularly in music. Broadcast radio is increasingly controlled by smaller numbers of media conglomerates (such as Cox, Jefferson-Pilot, and Bonneville). In some ways, this has led to more mainstreaming of the programming on broadcast radio, as stations often try to reach the largest possible audience to charge the highest possible rates to advertisers. Internet radio, on the other hand, offers the opportunity to expand the types of available programming.

There are two ways to deliver audio over the Internet: downloads or streaming media. In *downloads*, an audio file is stored on a user's computer. Compressed formats are the most popular form of audio downloads, but any type of audio file can be delivered through a web or FTP site. *Streaming audio* is not stored but only played. It is a continuous broadcast that works through three software packages: the encoder, the server, and the player. The *encoder* converts audio content into a streaming format, the *server* makes it available over the Internet, and the *player* retrieves the content.

3-D Digital Gaming

Advantages over the Analog Version

The terms *computer game*, *digital game*, and *console game* are used interchangeably. A computer game has two basic components: an output device and an input device. The output device is often a TV screen, computer monitor, or, in the case of portable games, a miniscreen. The output device enables the user to view the action of the game. Input devices are often more varied. For portable devices, they can be keyboards, and for desktop devices, they are usually a joystick or a controller.

A majority of the games released today share some common features:

- They are played on a console using a television set as an output device. The game software is accessed via a games console, to which input devices such as joysticks or controllers are attached, or the game software is accessed or downloaded via a satellite or digital-subscription-based system.
- They are played on a desktop computer (PC or Macintosh).
- They are played on small, portable games machines.
- They are played via consumer electronic devices, such as mobile phones and handheld PCs.

Games have been around since the dawn of humankind. The first games were played with sticks and stones; today, we have Monopoly, Risk, chess, and a host of others. So, why and how did digital gaming grow so fast and to such an enormous size (it was noted that in 2004, the electronic gaming industry was worth approximately $10 billion and growing at 25% to 40% per year)?

Most games are simulations of some aspect of reality. Video games have the ability to simulate this better than their older physical counterparts. Although the game *Life* simulates our trek through life's journey, its realism and ability to capture the mind are a testament to the imagination of its creators. Video games provide this realism and require less creative imagination. There are a couple of genres of simulations. The first is reality-based simulations. These include car racing games, business simulations, sports, combat, flying, and civilization development games. This genre of games is not limited to just entertainment. The business sector has long used games and simulations for training staff in developing fiscal, economic, and trading skills. The military has long used simulation-based games in combat training, and the health/medical sector is increasingly using similar realism techniques and technologies to those used in games. These simulations increase practicality while reducing costs. It is certainly cheaper and safer to train someone on a simulator than on a real jet aircraft.

The second genre of computer games involves adventure and fantasy. These games, with realistic graphics and physics-based effects, are based more on imagination and fantasy than reality-based games.

The third genre of games consists of mind stimulation games, such as Tetris, and conversions of traditional games such as Scrabble, chess, and Monopoly.

The common theme of these genres of games is escapism. The temporary removal of oneself from reality has been long sought after and is probably a fundamental human need.

The sales analysis and use of games have been well documented in many surveys and reports. The gaming industry, in its entirety, has become comparable in revenue, customers, and employees to the film and music industries, which generates approximately $18 billion annually.

Computer games are certainly becoming more pervasive in our world. Almost all consumer electronic devices (phones, PDAs, etc.) come with games already installed, and gaming consoles are becoming common in hotel rooms and airline seats.

Technical Differences between the Digital and Analog Forms

Perhaps the first difference is the manufacturers. Rather than Milton Bradley, it is the electronics companies that produce electronic games. As of today, there are three main manufacturers of digital gaming products. The three big players in this field produce different types of games, and most reviews of the three company products are unable to decide on a clear *winner*, instead concluding that each is more appropriate for a particular gaming group. The increasing dominance of the market by these three companies makes it difficult to see other manufacturers developing and sustaining sales for digital gaming products. The big three are as follows:

Microsoft—In late 2001, Microsoft launched its career in gaming with the X-box, followed by the X-box 360 (Exhibit 15.9) in 2005. Its latest console, the X-box One, was released in 2013.

Nintendo—In 2001, Nintendo launched the GameCube in Japan and the United States as a successor to its N64 console. In 2006, the Nintendo Wii was released and has sold over 100,000,000 units.

Sony—The Playstation3 (or PS3) has sold over 80 million units worldwide. The PS3 is the successor to the PS2, which sold approximately 20,000,000 units between its release in 2000 to 2012.

Many people also use their PC as a games machine (Exhibit 15.10a–d). Most PC users at least play with simple games, such as Minesweeper and Solitaire;

Exhibit 15.9 X-box.

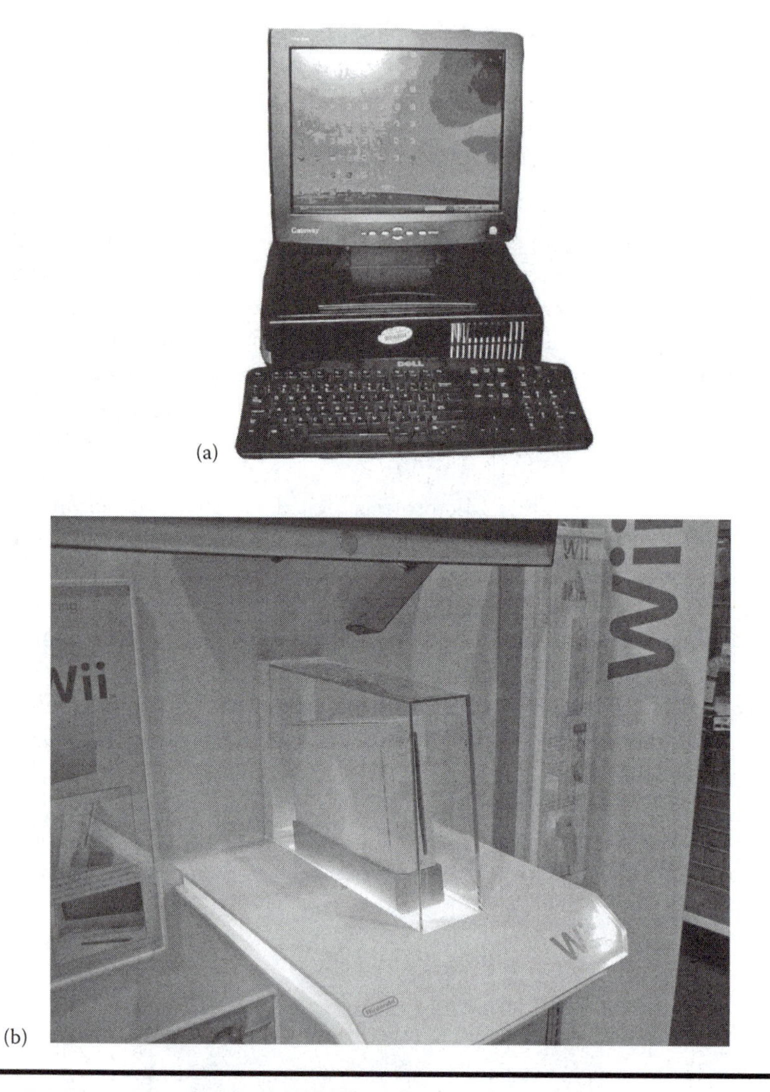

Exhibit 15.10 (a) PC and (b) Nintendo Wii.

(Continued)

however, many recent PC-based titles are of a quality and complexity that match that of leading console-based titles. Online games (especially combat-oriented simulations), civilization-building games, business tycoon simulations, and flight simulators are genres that are particularly strong on the PC.

Why use an expensive and multifunctional PC as a gaming console? Probably the primary reason is convenience and cost. Most people have a PC already, and

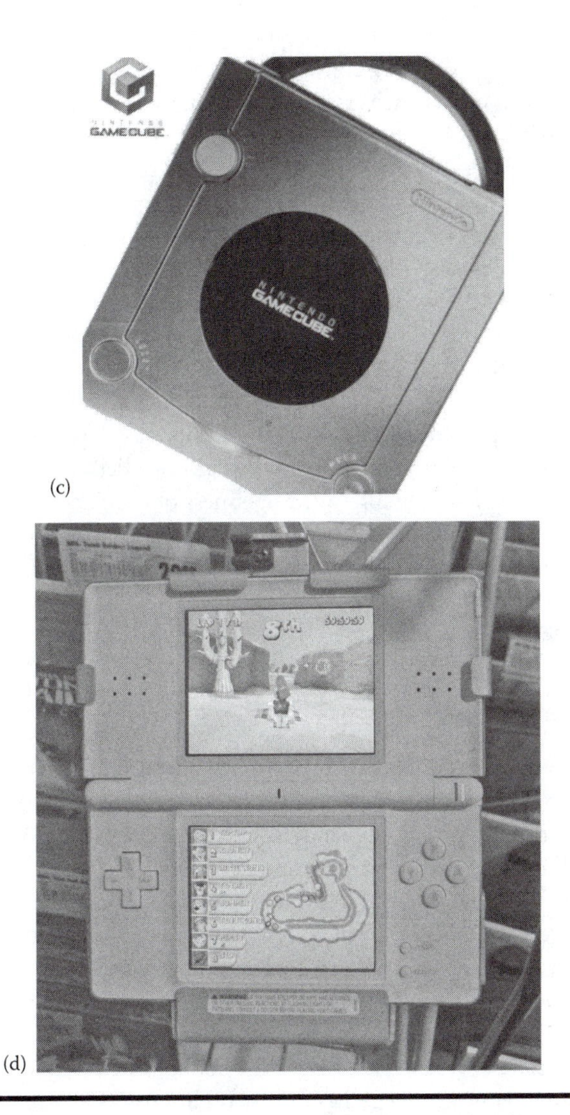

Exhibit 15.10 (Continued) (c) Nintendo GameCube and (d) Nintendo DS.

because it is paid for and already on your desktop, why not use it for games? The advantages of a gaming console follow:

Ease of system operation—Because gaming consoles are single-purpose devices, they are easier to use. Games consoles have very few buttons; software operation consists of inserting the cartridge and disc, and turning the power on. Usually, the underlying operating system is kept largely hidden from the user.

Fixed specification—Gaming consoles are fixed-specification devices; the graphics, internal memory, drivers, and so forth are identical across all units of a particular model.

Operational stability—PC operating systems and software often *crash* or *lock up*. Owing to the aforementioned fixed specification, this very rarely occurs with games consoles.

Openness of development—It is relatively easy to develop software or applications for the PC using a wide variety of tools and programming languages. Games consoles are a more closed environment, and expensive software and hardware are usually required to create the end product.

Range of peripherals—PCs come with a keyboard and mouse as standard; peripherals such as printers, scanners, and additional drives are commonplace. A games console typically comes with just a controller.

Range of applications—Owing to the openness of development, the PC can run a vast array of applications, including games, business applications (word processors, spreadsheets, and databases), educational software, Internet software such as web browsers and e-mail readers, and so forth. Games consoles run just games.

New Opportunities

Academic course work—As we saw in the early part of this chapter, digital media courses are growing in number, especially the field of gaming. The number of gaming-related courses has grown tremendously in recent years. Methodology in teaching gaming-related topics differs; some courses are tailored toward game design and programming, whereas others offer a more generic computer science qualification comprising several games-related modules. Also, research is beginning to show that gaming is a very effective teaching tool when properly utilized.

Gaming devices as entertainment—Gaming devices are not yet a common fixture in libraries, though they are starting to appear as an addition to existing entertainment media such as CDs, videos, and the fiction (book) section. More common is the provision of educational games for use on library PCs.

Convergence—Games are appearing on a number of electronic devices, in part to make such devices more commercially attractive to games players, and in part to increase the number of functions offered by the device.

Interactive TV—Unlike many technologies that have failed to fulfill predictions of rapid growth, interactive TV seems to be taking off in territories such as the United Kingdom. This medium is well suited to games; possibilities exist for people to, for example, play associated games running in parallel with television programs, such as quiz shows. The use of a television screen means that game developers are not faced with the restrictions of screen size and format that hinder mobile phones and other portable technologies.

Desire for participation—According to Ziff-Davis, the average consumer is switching viewing preference from TV watching to gaming. The desire of the consumer seems to be changing from being a passive participant to being an active participant.

Jobs—Recent studies have shown that the gaming industry currently supports a large number of full-time jobs in the United States. Predictions are for the industry to grow by over 75% in the near future. These jobs are highly skilled and well paid. A typical entry-level position will pay over $50,000. Not only the gaming industry hires graduates, but also there are several peripheral industries that support the gaming industry and also hire competitively.

Creation of master's degree programs—As mentioned before, there are numerous bachelor's degree programs in digital media and also many in digital gaming. Master's degree programs in the digital gaming field are beginning to come online. Michigan State University, Carnegie Mellon, University of Southern California, and Georgia Institute of Technology have recently launched master's degree programs and expect many others to come online soon.

Summary

In this chapter, we looked at the world of digital media and focused on the creation of the message and not just its transfer. Certainly, digital media has taken a strong foothold in today's world and promises, via convergence of technologies, to further strengthen its position.

We looked at four digital media sources: digital photography, digital video and digital audio, digital radio, and digital gaming. We looked at why digital photography has become such a strong leader in the digital media world, and some of the differences and similarities between digital photography and *wet* photography.

We compared digital video to digital photography and analog filming. Digital video, with its convenient filming and storage, has created a whole new type of vacationing experience and a brand-new web presence, for example, YouTube.

Digital audio has not become as strongly entrenched as manufacturers would like it to be, although it has created a robust economy in personal devices and the sale of digital music over the web.

Digital radio is in its infancy but promises strong growth. The key players are identifying and creating niche markets for their products, and, similar to a garage door opener, once you have tried it, it is hard to go back.

We ended this chapter with digital gaming, which is characterized by strong market penetration, innovative technology, and the fundamental human need for escapism. Simulations hope to expand this field beyond the confines of just entertainment and into education and training.

Review Questions

True/False

1. Digital media depends on the electronic age to flourish.
   a. True
   b. False
2. Analog photography is called dry photography.
   a. True
   b. False
3. Type 2 and Type 4 are the two different sizes of a CF card.
   a. True
   b. False
4. The term picture sharing has arisen owing to analog photography.
   a. True
   b. False
5. With satellite radio, it is possible to listen to the same station from California to New York.
   a. True
   b. False
6. Downloading and streaming media are the two ways to deliver audio over the Internet.
   a. True
   b. False
7. PC games are easier to use than gaming consoles.
   a. True
   b. False
8. The two types of memory cells used in CF memory cards are SLC and MLC.
   a. True
   b. False
9. MIDI is a new interface in digital media.
   a. True
   b. False
10. Dpi is a property of digital media.
    a. True
    b. False

Multiple Choice

1. Whose work was the turning point for digital logic?
   a. Ben Hake
   b. Francis Bacon
   c. Andrew Wildman
   d. All of the above
2. Cyan, magenta, and yellow together produce what?
   a. Black
   b. Red
   c. White
   d. Gray
3. What does DPI stand for?
   a. Digital Policy Institute
   b. Digital Photography Inc.
   c. Dots per Inch
   d. Digital Printing Inc.
4. What are some of the methods to make up for the inherent flaws of digital music?
   a. Oversampling
   b. Upsampling
   c. Downsampling
   d. All of the above
5. Which of the following companies were granted the license to broadcast satellite radio in 1997?
   a. XM
   b. Pandora
   c. Spotify
   d. All of the above
6. What is the difference between CF cards?
   a. Cost
   b. Storage capacity
   c. Speed
   d. All of the above
7. Which of the following are the transparent ribbon colors used in dye sublimation?
   a. Yellow
   b. Black
   c. Green
   d. All of the above

8. To achieve better quality, what do better photo printers add (mention two items)?
   a. Light magenta and light cyan
   b. White and gray
   c. Pixels per inch
   d. All of the above
9. What is the FireWire port also known as?
   a. Sony i-Link
   b. IEEE 1394 Port
   c. Plate
   d. S-Video
10. Which one of the following is a new interface in digital audio?
   a. XM
   b. DPI
   c. CF
   d. MIDI

Short Answer Questions

1. List the four most popular sources of digital media.
2. What are the additive and subtractive filters in a camera?
3. What is the main performance characteristic of a CF card?
4. List the three technologies used for printing.
5. What are the three major components of satellite radio, from a technology standpoint?

Chapter 16

Network Security and Management

Introduction

Network management and especially network security are the hottest areas and issues in the ICT field these days. With the consistent attacks by hackers to our corporate networks and our own personal devices, it appears that these two areas will continue to remain in the forefront of the ICT field for the foreseeable future.

In this chapter, we will cover network security and management. We will start off by discussing the types of attacks and explain how each attack affects our networks and computing devices. Next we will move on to some basic concepts in the security field—accounting, authorization, and encryption. We will follow with a discussion of the various tools, technologies, and strategies that can be employed to assist in securing our networks and computing devices against the bad guys that are out there. We also will discuss the benefits and payback of security.

Following this section, we will discuss network management—how to effectively and efficiently manage a midsize network. We will discuss the technologies and tools of network management and follow this up by best practices in the field. We close this section with a look at emerging bring your own device (BYOD) environments and their threats to security.

Although the elements discussed within are only a brief overview of many in-depth topics, they will give you a foundation in which to explore both network security and management. However, it is important to continually further and update your education in these arenas, as the landscape of network security is constantly changing.

Network Security Threats

Network security: Hard to live with it, impossible to live without it. Security is a balance. On one side of the scale is convenience. All users want this convenience (in the form of easy access, no passwords, always available). On the other side of the scale is security: the need for an organization, or an individual, to protect its assets, protect its secrets, and protect its equipment. The *right* security is one that forms a balance on this scale that is acceptable to both the users and the organization.

In order to convince users that security is important to them, a typical strategy is to note the amount of scams and crimes that are perpetrated every year. According to IC3, the 2012 Internet Crime report, an organization that monitors network and computer crimes, all computer security is at risk. The report cites that, in 2012, more than 289,874 complaints were filed. These complaints had an adjusted dollar impact of over $525 million. Of the complaints with monetary loss, the average dollar loss of those reporting was $4500. The majority of the complainants were male and from the United States (highest states were California, Florida, Texas, and New York). The popular crimes were auto fraud, impersonation, intimidation/extortion scams, and real estate fraud. We will learn in a bit that most of these crimes are in the category of *social engineering*.

In Exhibit 16.1 (Identity Theft Resource Center, 2014), you will find the breach statistics for 2005–2013. This provides the number of corporations that were successfully broken into by hackers. You will find that the business and health-care fields are the most broken into sectors, as they have the most information to be gleaned. Hacking is the most used method (we will define all these categories later) with subcontractor failures rising fast. Do remember that each one of these breaches represents hundreds if not thousands of customer records.

Who Are the Bad Guys?

Who are these attackers? Why do they do it? What do they get out of it? Before we get started in discussing the types of attacks, we will start with a profile of different attackers. Computer attackers fall into general categories of hackers.

Hackers

Hackers are usually young and want to beat the system by breaking into networks of computer systems. They are intellectually acute and sometimes morally challenged. A hacker is a person who enjoys the technical field, exploring computer systems and networks as a challenge. Some hackers are driven by sharing, openness, decentralization, free access to computers, and even the goal of world improvement. Others are motivated by financial/personal gain or more destructive goals. The field of hacking can be further broken down into different categories (white hats, black hats, gray hats, and elite hackers, to name a few).

ITRC Breach Statistics: 2005 - 2013

Industry Sector	# of Breaches 2005	# of Breaches 2006	# of Breaches 2007	# of Breaches 2008	# of breaches 2009	# of breaches 2010	# of breaches 2011	NUMBER of breaches 2012	PERCENT of total breaches 2012	NUMBER of breaches 2013	PERCENT of total breaches 2013
Business	28	69	130	243	208	279	198	172	36.4%	211	34.4%
Educational	75	80	111	131	78	65	60	65	13.7%	55	9.0%
Government/Military	21	98	110	110	90	104	48	53	11.2%	56	9.1%
Health/Medical	13	43	64	94	65	160	87	165	34.9%	269	43.8%
Financial/Credit	20	31	31	78	57	54	28	18	3.8%	23	3.7%
	157	321	446	656	498	662	421	473		614	
Category			# of breaches 2007	# of breaches 2008	# of breaches 2009	# of breaches 2010	# of breaches 2011	NUMBER of breaches 2012	PERCENT of total breaches 2012	NUMBER of breaches 2013	PERCENT of total breaches 2013
Insider Theft			27	103	85	102	56	40	8.5%	72	11.7%
Hacking			63	91	97	113	108	129	27.3%	160	26.1%
Data on the Move			124	137	78	110	76	57	12.1%	80	13.0%
Accidental Exposure			90	95	59	71	44	41	8.7%	46	7.5%
Subcontractor			51	68	37	58	32	53	11.2%	88	14.3%
Employee Negligence								33	7.0%	57	9.3%

Identity Theft Resource Center, Copyright 2014

Exhibit 16.1 Types and numbers of security breaches.

A *white hat* is a person who does penetration testing on a system in order to find its weaknesses and fix them. Often employed by software companies, these personnel are doing this for non-malicious reasons and are often called ethical hackers.

Black hat hackers are performing break-ins for nefarious purposes and are violating all computer security policies. They want to steal data, destroy data, make the network unusable, or modify the data on the machine. These are the people who steal credit card numbers and sell them on the black market.

In between the white hat and black hat hackers are the *gray hats* who break into computers, inform the owner of the flaw in their system, and offer to fix it for a fee.

Elite hackers are a social group of the pros in the field. The Masters of Deception is a group of these elite hackers who find, create, and share weaknesses within computer and network systems.

Types of Network and Computer Attacks

There are as many types of attack strategies as there are attackers. Most attackers adapt and put a signature on their style of attack. But attacks can generally be categorized into some basic and broad categories. Here we will discuss social engineering attacks, denial of service (DOS) attacks, malware, and sniffing.

Social Engineering Attacks

Social engineering attacks are low tech and simply prey on the kindness and habits of fellow humans. A social engineering attack can consist of

- Calling up a bank, pretending to be the account owner in order to get their account number and password
- Walking into a store and assuming another person's identity
- Dumpster-diving into individuals' and stores' garbage looking for credit card numbers and personal information
- Shoulder surfing; looking over a person's shoulder, or keyboard, for a password

In social engineering attacks, it is usually a person dealing with another human being, although current *phishing* scams and *social media* scams do make use of technology to dupe fellow humans.

DOS Attacks

In DOS attacks, the goal is to deny the computer or network use to its owner. Similar to the old day *union picket line*, the goal is to disrupt service to send a message. The DOS attacks are typically the work of black hat hackers. DOS attacks come in many flavors such as

- Requesting a computer to determine the value of pi to the 10,000,000th digit, therefore using up its resources so it cannot do work for its owner.
- Sending innumerable requests to an e-mail server so that the server cannot effectively process e-mail for its legitimate users.
- Sending innumerable routing updates through the network, therefore effectively bringing the network to its knees and stopping the flow of meaningful and useful traffic.
- Distributed DOS (DDOS) attacks, where *zombies* are first recruited (some software is unknowingly put onto your computer or personal device) and then later called into action by the black hat hacker. The action requested of the zombie can be sending of e-mails to a target computer to overload the computer or the network, or requests for information that keep the device busy and therefore not able to handle serious requests.

Malware Attacks

With malware attacks, the attackers do not steal anything from your computer, or network, but instead put something into your system. There are three different types of malware that attackers can put on your network or computing device: viruses, worms, and Trojan horses.

Virus

A virus is a piece of computer code that has the intention to replicate itself from machine to machine. A virus attaches itself to a program or file so it can spread. It infects as it travels around the network. Fortunately, viruses need human action to spread and this is where it can be stopped. Viruses usually come in e-mail attachments, and the act of opening the attachment will activate the virus. Viruses can be innocent, popping up annoying comments to distract us, or can be destructive by erasing your whole hard disk.

Worm

A worm is a type of virus that replicates by itself. Worms sometimes carry a payload that could read your whole e-mail directory and send itself to everybody on your e-mail list. Worms do not take human intervention to replicate and because of this can quickly and efficiently spread through a total network very expeditiously.

Trojan Horse

A Trojan horse is a program that appears useful but has dishonorable intentions. Typically, a Trojan horse will appear as a program to help you, such as an e-mail sorter, or a nifty toolbar, but, when opened, will contain a worm, virus, or another malware package.

Sniffing

The intent of sniffing is to listen to your network or computing device for interesting information, such as passwords, usernames, and other personal information. This information is then used for nefarious purposes. To sniff or eavesdrop on a network is not a very difficult process and can lead to tremendous results. So, what can be done with information one hears on the network?

Data Theft

Sensitive information about sales figures and social security numbers gets stolen off the wire and is used for nefarious purposes. Additionally, this sensitive information is not forwarded to the original sender, therefore denying him/her of the information.

Data Alternation

While listening on the wire, one can make changes in the information. For example, you deposit $1000 into your bank account, but the information is intercepted and altered so only $100 goes into your account and $900 goes into the hackers account.

Spoofing

After one sniffs the wire, they can see the password and username being sent. Subsequently, they later use this password and the username to log on to your bank account to transfer money from you to them.

Knowing the attacker's methods, and understanding the attacker's strategies and motivations, puts the defenders in a decidedly advantageous position.

The Payback of Security

Maintaining network and computer security takes a lot of human effort and a lot of money (firewalls, security devices, personnel). Is this effort and money well spent? Is it worth the effort? What does it provide to the organization? Below are the four main reasons that security is justified: customer trust, mobility, increased productivity, and reduced cost.

Customer Trust

A strong security stance assures customers that sensitive information, such as credit card numbers or confidential business details, will not be accessed and exploited. Business partners will feel more confident sharing data such as sales forecasts or prerelease product plans. Without this trust, customers will leave and other companies will not do business with you. Additionally, security can give your partners access to information on your network, helping you collaborate and work together more effectively.

Mobility

Strong security tools and policies allow employees to work remotely and on the road. This access is convenient, moral boosting, effective, and efficient for both the employee and the company.

Increased Productivity

By decreasing the amount of spam, viruses, and other security threats, your employees can concentrate on their work task more effectively and efficiently.

Reduced Cost

If your network or computer system is down, how much does it cost the company? How will people go to your site to buy items? The cost of *being down* for some companies is very large, and the cost of avoiding that with proper network security is well worth the cost. Being proactive with security is far more cost-effective than cleaning up afterward. There are many untold stories of companies that folded when they lost their data and computer systems.

Basics of Security

With knowledge of the types of attackers and types of attacks understood, we will now move onto some of the basic elements of security. There are three key elements in the implementation of security precautions: authentication, accounting, and encryption. These three elements form the basis for most security rules and procedures and will be covered next.

Authentication

Authentication, sometimes referred to as authorization, has as its goal to allow only the correct entity to use the correct resources. For example, if you have the key to your car, you can use the vehicle and therefore you are authorized. Whoever you give the key to therefore has the authorization or authentication to use the car. There are three basic strategies to network and computer authentication, and each has an increasing level of complexity and security.

What You Know

This strategy is the simplest form of authentication and involves something you know, such as a password. The password, your name, a number, your mother's maiden name, or the passcode to your car can give you access to your computer, your account, or your car. One key to this strategy is not allowing anyone else to know what you know, your password, by not writing it down or telling anyone. The other key is for the user to select a password that will be easily remembered but not easily determined. For example, a birthdate (Oct1954) is easily remembered but also easily determined. A daughter's name mixed with her birthday is easily remembered but hard to determine (for example, Ka0988tie). Passwords are being required that demand to have at least one number (5), to have one lower and UPPER case character, to be of a certain length, and to have one special character (*). This requirement is necessary for your security but demands some thinking on your part to find an easy-to-remember password.

What You Have

Going one step further than the previous strategy, this involves not only what you know but also what you have in your possession. For example, you know your house number and street and you have the key. Therefore, these two elements allow you to be authorized to use the house. The ICT field uses devices such as a *smartcard* for the employee to have with them. The employee knows their username and has the smartcard, which gives him/her the password for a specific time period. Without both of these elements, access to the account would not be possible.

Who You Are

The strategy relies on authentication due to some biometric, unique, characteristic of a person. Currently, voice prints, fingerprints, retinal scans, and DNA are being used for biometric authentication. Certainly the highest level of security authentication comes with a higher cost, a higher inconvenience (Exhibit 16.2).

Exhibit 16.2 Picture of a biometric device.

Accounting

Accounting is the art and science of watching who did what, when, how, and possibly why. Essentially accounting is watching to see if your network or computer is being probed or tampered with or being improperly used. For example, (1) did someone try to hack into your network using a password generator? (2) Where did this attack come from? (3) Why did the engineering department request information on salaries? (4) Who was using the network last night at 4:30 AM? What were they doing?

Although not a glamorous part of security, without an accounting trail, and someone to scrutinize the breadcrumb trail, you have no chance of catching a perpetrator or stopping one who is persistent.

Encryption

Encryption is the process of taking a readable message and transforming it to an unreadable format during transit and storage. In essence, only the sender and the receiver can interpret or read the message properly. This concept is important in security, as we often pass key information in e-mails, spreadsheets, and telephone calls. Information such as passwords, sales projections, and financial account information can be key and if read by black hats could be used against us (Exhibit 16.3).

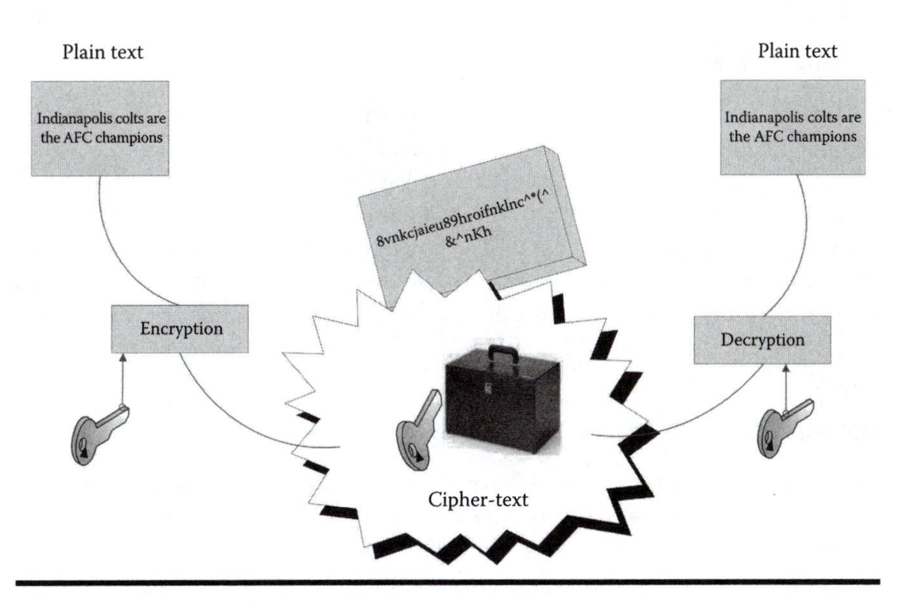

Exhibit 16.3 Encryption process.

The main tool to encrypt messages is called an encryption algorithm. An algorithm is a set of defined tasks designed to produce an end result, in this case an encrypted message. For example, if we take every letter in the alphabet and add five letters to it, the message *hello* would be transmitted as *mjqqt*. The receiver, knowing the algorithm, and the five letters, could reverse engineer the code back to the original *hello*. There are numerous algorithms available.

Some are extremely hard to break; others are easy. Some of the algorithms are only for government use, others for private use (Exhibit 16.4).

Take for example the advanced encryption standard (AES). This particular algorithm is fairly hard to break and is the base standard for US government organizations. The algorithm is considered a substitution-permutation type, which implies substitution of certain elements (usually the keys) and permutation of other elements. AES uses a 4 × 4 matrix of bytes; it then substitutes values for a similar 4 × 4 matrix. By repeating these substitutions/permutations by a number of cycles, your plaintext is transformed into cipher-text that is unreadable. Each cycle of processing contains stages that include your encryption key. A reverse cycle is applied to bring your cipher-text back into the original plaintext using the same encryption key.

An important concept and difficulty in encryption is the type and the transfer of the *key*. In our previous example, the key was the five letters used to encrypt *hello*. If the key is compromised (in either its algorithm or in the transfer) and read by those who we do not want to give access, the message is compromised. So the task comes down to, "how do I create a good key and get the key to the intended receiver in an effective, efficient, and totally secure manner?" There are a few basic ways to do this.

Symmetric Key

Symmetric key encryption uses two keys at sending and receiving site that are textually related, which means that they could be identical or slightly variant. Although symmetric keys are totally secure, having similar keys at both ends compromises the integrity of the key and therefore the communication. This demands the frequent changing of keys (called key management) and is a relatively heavy burden placed on symmetric key encryption. Symmetric keys must be distributed by either face-to-face meetings, use of a trusted courier, or sending the key through an existing encryption channel.

Asymmetric Key

With asymmetric key encryption, sometimes called public key encryption, there is a public and a private key. This type of cryptography allows users to communicate securely without having prior access to a shared secret key. This is accomplished by using a pair of keys (one called the *public key* and the other a *private key*), which are mathematically related. The data that were encrypted by the public key can only be

Name	Year developed	Type	Security level	Implementation	Key points	Speed
RSA	1977	Asymmetric	Military grade	2048 bit public key	• Public key common for many used for encrypting messages • Private key known only to selected users used for decrypting messages	Very slow
DES/3DES(Data Encryption Standard)	1977	Symmetric	Low	40-56 bit shared secret	• It's a Block cipher algorithm where a fixed length of text is encrypted • Each block is 64 bits long, 56 bits of encrypted text and the remaining is for parity check • Due to recent developments some experts consider DES no longer effective or safe against attacks	Fast
BlowFish	1993	Symmetric	Military grade	256-448 bit shared secret	anywhere from 32 bits to 448 bits. • As 64 bit block size is too short encrypting more than 232 data blocks can begin to leak information about the plaintext	Fastest
MD5(Message Digest Algorithm 5)	1991	Asymmetric	High	128 bit message digest	Operates on 64 bit blocks and has a 128 bit key • Works by interleaving operations from different groups like modular addition and multiplication, and bitwise XOR	Slow
AES (Advanced Encryption Standard)	2001	Symmetric	High	128, 192 or 256 bit key	Very secure as it is computationally infeasible to find a message that corresponds to a given message digest, • Any change to a cipher text will result in a different message	Fast

Exhibit 16.4 Popular encryption algorithms.

decrypted by the private key. In asymmetric key encryption, the private key is kept secret, while the public key may be widely distributed.

An analogy for asymmetric keys that is often used is that of a locked store front door with a mail slot. The mail slot is exposed and accessible to the public; its location (the street address) is in essence the public key. Anyone knowing the street address can go to the door and drop a written message through the slot. However, only the person who possesses the matching private key, the store owner in this case, can open the door and read the message.

The term *asymmetric key cryptography* is a synonym for public key cryptography though a somewhat misleading one. There are asymmetric key encryption algorithms that do not have the public key–private key property noted above. For these algorithms, both keys must be kept secret; that is, both are private keys.

There are many forms of public key cryptography:

- Public key encryption—keeping a message secret from anyone who does not possess a specific private key
- Public key digital signature—allowing anyone to verify that a message was created with a specific private key
- Key agreement—generally, allowing two parties that may not initially share a secret key to agree on one

Asymmetric key encryption is usually much more computationally intensive than symmetric encryption, but the wise use of this technique enables a wide variety of applications. Exhibit 16.5 illustrates these encryption processes.

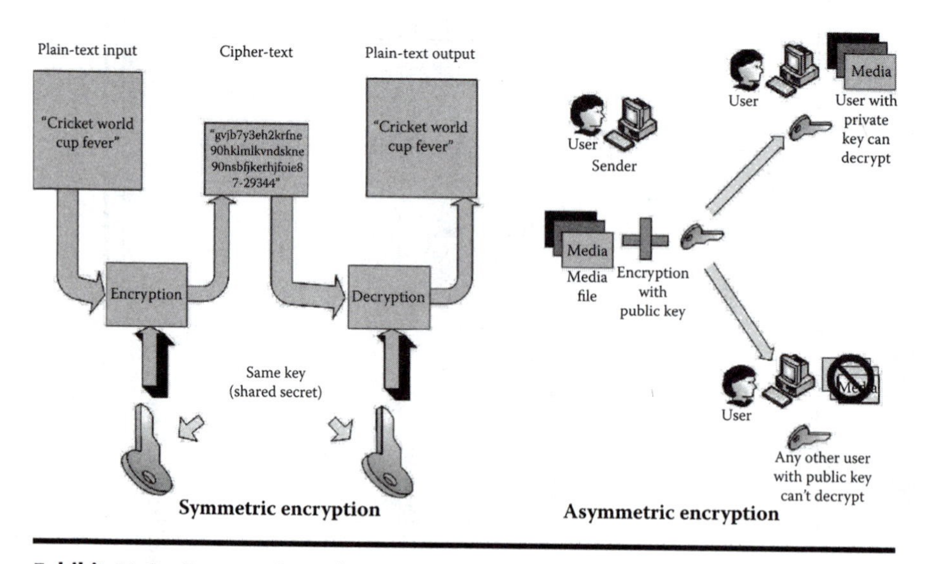

Exhibit 16.5 Symmetric and asymmetric key encryption.

Access Control and Site Security

Many people think that communication security is merely applying a set of electronic tools (Firewalls, IPSs, etc.) to a network. Certainly this is a part of it, but a very large part of security is simpler and more mundane: controlling access to physical locations and securing the cables. If a hacker could gain access to the wiring and other electronic elements of a network (routers, switches, and hubs), the results could be disastrous. For example

- *Wiring closets*: If access is obtained here, information could be changed or sniffed or the network could be totally disabled.
- *Wire port*: If access is gained to the wiring or an open port, a thief could sniff out password or other critical information, or information could be injected into the network.
- *Server/computer room*: If access is gained here, critical information could be modified as well as all systems disabled or compromised.

Key strategies in attenuating these access control issues are as follows:

- *Electronic door locks*: All communication rooms should have electronic door locks (see Exhibit 16.6) that are tied into a centralized database so access can be regulated (for a terminated or disgruntled employee) and audited (knowledge of who entered and leaves the room and when).
- *Protected server room*: When data are stored in a centralized location and 24 × 7 uptime is necessary, the server/computer room must be protected with video recording equipment and Halon fire extinguishing equipment. The video equipment inhibits tampering with the equipment plus allows a trail for accounting purposes. The Halon fire extinguishing equipment protects the data from fire.

Exhibit 16.6 Electronic door lock.

- *Port lock down*: Ports that are not used for normal connectivity should be disabled or put on a *dirty* network. A dirty network allows access to the Internet only and not internal resources.

Security Policy

A security policy, like access control, is not a technical item of security but is more than less *the* critical item of security. A security policy, from an ICT perspective, is like an architectural plan for a house. It provides the goals and objectives and specifies what will be allowed, what will be denied, and what to do it the event of a breach in security.

A security policy is a *written* document that decides and documents

- The technical architecture plan for the network
- The traffic that will be allowed into and through the network
- The traffic that will not be allowed into and through the network
- The procedures and process that happen when a breach in security exists
- The agreement employees need to make with the organization for use of its technology
- The location and criticality of organizational information
- The recovery plan when security is breached

Typically security plans are created by the ICT department in close cooperation with upper management and provide the directives in where to go in regard to security-related issues.

Network Security Tools and Techniques

We have covered the profiles of attackers (and their motivations), the types of attacks that can occur, and the basic elements of ICT security; it is now time to explore some of the tools of the trade. First we will start off with *network security*; second we will explore the tools or the trade for *host security*. Network security deals with the network and all of its associated pieces. Host security deals with the components, personal computers, servers, printers, PDAs, and other devices that the end users interface with.

Network Security

Depth in Design

A tool, or strategy, of proper network security is the depth-in-design concept (see Exhibit 16.7). As an analogy, an individual might protect his/her personal valuables with an outside fence, a front door lock, a bedroom door lock, an alarm system, and ultimately a safe. We put the most valuable items in the safe and the intruder has to

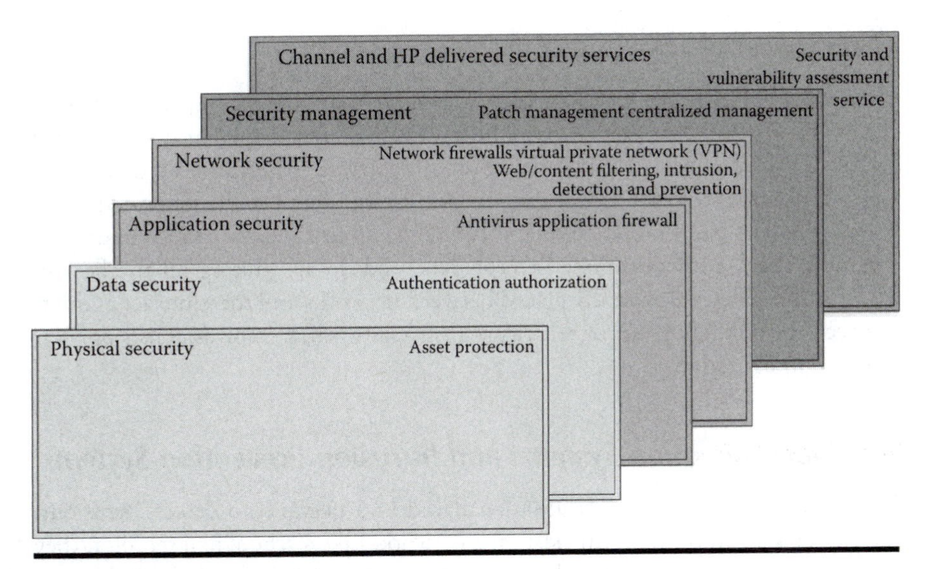

Exhibit 16.7 Layers of depth of security.

pass the many layers of security before he/she can get to these valuables. Similarly, a network needs to be designed in layers of depth with the most important information located at the core of the network. These layers typically are the external network, the demilitarized zone (DMZ), protected zone 1, and protected zone 2.

1. *External network*: The external network is the Internet, the public roads, and the community shared network that is totally unprotected and open to all. You place on this layer only what you can afford to lose or have corrupted.
2. *The DMZ*: One step more secure is the DMZ. In this zone, we place information that we want the public to see and get at easily but want to have a modicum of security on. Typically, a company will place its web server, e-mail client, FTP server, and other elements that can serve its outside image appropriately.
3. *Protected zone 1*: The protected zone is meant for employees, authorized guests, and other authorized parties of the organization. On this layer are many user accounts, corporate data elements, and partner information that need protection but also must be accessible by many authorized people.
4. *Protected zone 2*: On this most secure layer is information that needs to be well protected from outside parties and from unauthorized internal employees. Information such as payroll, accounts receivable, and other financial information are usually kept here.

Between layers or zones is a figurative door or gatekeeper that only allows access to deeper more secure zones with the right authentication. Accounting and movement at more secure layers are tracked closely. The following are some of the tools used to keep these layers secure and to monitor and track access and actions.

Firewalls

Firewalls are the devices used to keep information and people separate between the layers. Firewalls can come as specific hardware designed appliances or software products. For personal use, firewalls are usually software-based and can range from freeware to around $200. For corporate use, firewalls are usually dedicated hardware appliances (see Exhibit 16.8) and can range up to many thousands of dollars. Firewalls check authentication, encrypt data, and do accounting on the flow of information. They are typically placed between networks and therefore act as doors between the different layers of security within a network. Exhibit 16.9 lists popular firewalls in use today.

Intrusion Detection Systems and Intrusion Prevention Systems

Intrusion detection systems (IDSs) are placed in a network to detect intrusions, viruses, and break-ins that can exist on a network. Like a school hall monitor, they sit and watch what goes on, watching for signatures of malicious attacks. Signatures are like profiles that warn you of corporate stealing, viruses, or attacks. If an IDS senses a recognized signature from a virus, or a DOS attack, it warns the system administrator by e-mail, pager, or phone call and waits for him or her to respond.

Intrusion prevention systems (IPSs) take this concept one step farther. They not only detect the viruses, malware, or attacks but also prevent them from spreading throughout the network. For example, if an IPS detected a packet with a worm enclosed, it would automatically

- Quarantine the packet
- Warn the recipient and sender of the violation
- Notify the system administrator of the potential attack
- Attempt to fix the worm or delete it

Exhibit 16.8 Firewall hardware appliance.

Products	*Features*	*Intruder/hacker/ detection tools*	*Internet tools*	*Setup and management*	*Help/support*	*Supported configurations*	*Type*	*Price*
Zone alarm	• E-mail protection • File protection • Personal information protection • Registry protection • Port monitoring • Network traffic monitor • Data filtering	• Intruder alert • Intruder ID lookup • Intruder tracking log	• Stealth mode • Pop-up blocking • Cookie blocking • Spyware blocking • Trusted website list • Blocked website list • Website history list	• Password protection • Individual user settings • Network time restrictions • Automatic software rules • Instantly disable firewall • Instantly block all traffic	• Phone support • Live chat • E-mail/online forms • Easy upgrades	• Windows 8 • Windows 7 • Vista/XP	• Software	Free
A shampoo Firewall	• E-mail protection • File protection • Personal information protection • Registry protection • Port monitoring • Network traffic monitor • Data filtering	• Intruder alert • Intruder ID lookup • Intruder tracking log	• Stealth mode • Pop-up blocking • Cookie blocking • Spyware blocking • Trusted website list • Blocked website list • Website history list	• Password protection • Individual user settings • Network time restrictions • Automatic software rules • Instantly disable firewall • Instantly block all traffic	• Phone support • Live chat • E-mail/online forms • Easy upgrades	• Windows 8 • Windows 7 • Vista/XP	• Software	Free
Norton Firewall	• E-mail protection • File protection • Personal information protection • Registry protection • Port monitoring • Network traffic monitor • Data filtering	• Intruder alert • Intruder ID lookup • Intruder tracking log	• Stealth mode • Pop-up blocking • Cookie blocking • Spyware blocking • Trusted website list • Blocked website list • Website history list	• Password protection • Individual user settings • Network time restrictions • Automatic software rules • Instantly disable firewall • Instantly block all traffic	• Phone support • Live chat • E-mail/online forms • Easy upgrades	• Windows 8 • Windows 7 • Vista/XP	• Software	Free
PC Tools	• E-mail protection • File protection • Personal information protection • Registry protection • Port monitoring • Network traffic monitor • Data filtering	• Intruder alert • Intruder ID lookup • Intruder tracking log	• Stealth mode • Pop-up blocking • Cookie blocking • Spyware blocking • Trusted website list • Blocked website list • Website history list	• Password protection • Individual user settings • Network time restrictions • Automatic software rules • Instantly disable firewall • Instantly block all traffic	• Phone support • Live chat • E-mail/online forms • Easy upgrades	• Windows 8 • Windows 7 • Vista/XP	• Software	Free

Exhibit 16.9 Popular firewalls.

IDSs and IPSs sometimes make mistakes. Some mistakes are very bad and others less so. A false positive, or *Type 1 error*, occurs when the device reports certain activity as being malicious when it is not. This wakes up the network director and causes a panic when one is not warranted. The other type of error is a false negative or *Type 2 error*. It occurs when the device does not detect actual malicious activity and fails to raise an alarm when it is needed.

Honey Pot

As the name suggests, a honey pot is a device that attracts hackers. It appears as a computer that is running an old and vulnerable operating system (OS), has many open ports, and no password protection. To hackers, these honey pots appear to be easy to access targets. In reality, the device is well monitored for what attacks take place on it, with it, and by whom. The information gained is critically important in understanding what tools and methodologies attackers use and where the attacks are coming from. With this information, the network and the computers can be better prepared to ward off attacks. Be aware that honey pots are controversial items in that they can be perceived as entrapment.

Host Security

In this section, we will cover the other side of ICT security, that being host security. As previously noted, there are two sides to security: network security and host security. Both must have attention paid to them in order to secure the data within an organization. If both are not covered, data security is at risk. *Host* in this case is referring to personal computers, laptops, servers, PDAs, iPods, and even smartphones.

OS Security

The first element that must be secured in a host is the OS. The OS (e.g., Mac OS, Android, UNIX, Linux, and Microsoft) must be secured as a front line of defense against host attacks. Hackers look for vulnerabilities in the OS in order to perpetrate an attack. The easiest way to secure the OS is to keep it up to date. OS manufacturers are constantly finding and plugging holes and reacting to attacks on their products. These manufacturers are constantly putting out updates with patches to their OS, and these patches must be installed on the hosts. Daily new updates should be allowed or searched for and installed.

Software Security

Similar to OS security, software (application) security must also be maintained. Patches and fixes must be installed for applications like Internet Explorer, Microsoft

Office, and other popular applications. Security breaches can easily be found in these applications and can result in loss of integrity, loss of passwords, and even a total loss of data. Similar to OS security, manufacturers will create and release patches for their products. You are responsible for the installation of the patches.

Antivirus Programs

Even with up-to-date OS and application software, viruses and other malware can get into a host. To thwart these attempts, a program must be installed on each host to look for these malware programs and stop them before they enter the host. This is the job of antivirus programs such as those shown in Exhibit 16.10.

Products	*Features*	*Scanning capabilities*	*Updates*	*Other features*	*Supported configurations*	*Prices*
Bitdefender Total Security	• WCL level 1 certified • WCL level 2 certified • Virus definitions updated hourly	• On-access scanning • Real-time scanning • On-demand scanning • Scheduled scanning • Heuristics scanning • Manual scanning	• Automatic definition and program updates • Manual definition and program updates	• Bitdefender Safepay™ • Keeps hackers at bay by automatically opening all your e-banking and e-shopping pages in a separate, secure browser • Block unwanted e-mails	• Windows 8 • Windows 7 • Vista/XP	$59.99
Symantec Norton 360	• WCL level 1 certified • WCL level 2 certified • Virus definitions updated hourly	• On-access scanning • Real-time scanning • On-demand scanning • Scheduled-scanning • Heuristics scanning • Manual scanning • Adware/spyware scanning	• Automatic definition and program updates • Manual definition and program updates	• SONAR Behavioral Protection proactively detects dangerous files • Network mapping and monitoring • Built-in intelligence	• Windows 8 • Windows 7 • Vista/XP	$49.99
Microsoft Security Essentials	• WCL level 1 certified • WCL level 2 certified • Virus definitions updated hourly	• On-access scanning • Real-time scanning • On-demand scanning • Scheduled-scanning • Heuristics scanning • Manual scanning	• Automatic definition and program updates • Manual definition and program updates	• Runs quietly without hurting PC performance • Cloud-based	• Windows 8 • Windows 7 • Vista/XP	Free
McAfee LiveSafe	• WCL level 1 certified • WCL level 2 certified • Virus definitions updated hourly	• On-access scanning • Real-time scanning • On-demand scanning • Scheduled-scanning • Heuristics scanning • Manual scanning	• Automatic definition and program updates • Manual definition and program updates	• Protect all the PCs, Macs, tablets, and smartphones • Store personal data in an online vault-accessed only by face and voice recognition	• Windows 8 • Windows 7 • Vista/XP	$39.99

Exhibit 16.10 Popular antivirus programs (2013).

These antivirus programs look at incoming data (arriving through e-mails, IMs, and other programs) and then inspect each packet and compare the contents against known signature of malware. If a virus or worm is found, the packet is quarantined. New viruses and worms are constantly being created, and virus programs must be updated daily to keep up with the new attack schemes.

Spam/Pop-Up and Cookie Blockers

Although not a critical mechanism to destroying the integrity of data on hosts, spam/pop-ups and cookies can be very annoying to the user and an unproductive waste of one's time and effort. There are blockers for these items, some free, some with a cost that will

- *Stop pop-ups*—those annoying screens that pop up on our computer without warning or request
- *Inhibit spam*—stop most junk mail from entering your mail box
- *Inhibit cookies*—stop cookies, programs that load themselves onto your computer and keep track of your activities, from being allowed on your host

Network Administration and Management

The goal of network management and administration is to assure that the network is properly operated and maintained (to ensure consistent and high uptime) and that the integrity of the data is maintained.

Key Aspects of Network Administration and Management

Major components of network management and administration are as follows:

- *ICT auditing and assurance*: Ensuring the data on the network is kept to a high degree of integrity, that it is not stolen or changed and that standards are maintained. Most current organizations have internal data standards and most also have to follow, by law, some external standards, codes, or regulations. Some external standards are Health Insurance Portability and Accountability Act (HIPAA) and Sarbanes-Oxley.
- *Backup and disaster recovery planning*: Ensuring the data on the systems is backed up so that it can be restored after a disaster. Additionally, these steps assure that a business disaster recovery plan is in place so that the reaction

after a disaster (flood, hurricane, power outage, etc.) is predictable and will assure a smooth business operation into the future.

- *Service level agreement (SLA)*: A SLA is a contract with the organization, users, or outside contractor that describes the scope and reliability that will be maintained. Statements such as "The network will maintain a reliability of 99.999%" provides direction to the ICT staff as to the amount of downtime allowed and also to the user for their expectations of network performance. Detailed and legally binding SLAs are becoming the norm within and between organizations.
- *Sustainability*: For ICT to be considered a strategic business asset, it must be sustained long term. To do this, proper budgeting must be in place to replace old equipment, keep software current, assure connectivity with the outside world, and be properly staffed. ICT is not a one-shot deal in which money is invested once for a network and computer and then forgotten. It must be formally maintained, and budgeting is the method of doing that.
- *Management of personnel*: A network, and an ICT division, is made up of more than computers and electronic devices. These devices do not work without dedicated, trained, and professional people monitoring and managing them. These personnel, from the director to the technician, must also be managed. Management includes hiring, firing, training, personnel development, contract negotiations, and all the various items that it takes to manage people.

Network Management Tools

Like network security, there are tools to help manage the network. To understand more of the technical parts of network management, some of these tools and techniques will be presented next.

Client/Server Architecture

Although there are many different types of architectures in networks, the client/server architecture is very prevalent in today's world. The Internet is based on client/server architecture. But what is the difference between these two elements? What exactly are clients and servers?

Client

We have used the term client a lot in this book. To get more into depth, let us describe the client types and roles within the network. Clients can be broken down into three categories: (1) typical *desktop clients* with hard drives or disks, (2) *diskless clients*, and (3) *portable clients*.

As the name implies, a typical *desktop client* is your average personal computer that contains a hard drive for applications, data storage, and a full OS. A *diskless client* has no hard drive and therefore no applications in it or storage medium for data. A *diskless client* relies totally on a server in the cloud to be effectively used. Typically *diskless clients* are used where security is an important element of an organization. A *portable client* (laptop, tablet, mobile phone, etc.) is similar to a disk client, but its form factor is one where battery power, size, and weight are critical design criteria.

All of these types of clients have commonalities. As clients, they have four purposes:

- Accept user data and application commands
- Decide to process the command or direct it to the server
- Direct the command to the local OS or to the server
- Pass the data from the network to the OS or application running on the client

Server

A server is a computer designed to serve a multitude of users (unlike a client that typically serves only one user). To accomplish this task, a server requires a number of distinct characteristics:

- *More robust technology*: High-speed capability, high amounts of RAM, multiple CPUs, and redundancy built in.
- *Network OS (NOS)*: To serve multitudes of users, in a variety of ways (TCP ports), typically a server runs a robust OS called a network OS. NOSs have more robust network management tools, are designed to be used by a large number of clients, authenticate users, and provide access to shared resources.
- *Centralization of communication tasks*: Communication applications and tasks are often consolidated on a server and provide efficiency to an organization. Examples of such servers are (a) print server, (b) security authentication server, (c) firewall, (d) NAT server, and many others.

Network OSs

As noted before, NOSs enable communications between multiple devices and the sharing of information across a network. To operate, a NOS must have certain features:

- *Performance*: It must be able to provide fast internal performance at reading and writing data and must be able to communicate these data quickly over a communication network.

- *Management and monitoring*: A management interface must be available to monitor server performance, client administration, and disk storage.
- *Security*: A NOS must protect itself and the data it holds, so authentication, access control, and encryption services must be available.
- *Scalability*: A NOS must be scalable as new hardware is added and as new clients and users are added. This must be a seamless scalability and be able to be done *on the fly* in real time.
- *Fault tolerance*: Due to the many users relying on the network and its services, the NOS needs to be robust and resistant to faults. Dual power supplies, multiple processors, and a strong fault tolerance and authentication must be allowed for.

NOSs are built to provide services to hosts, clients, and users. Many of these services have come to be expected in a network, and it is the NOS that provides them. Some of these more common services are as follows:

- *File sharing*: Provides the ability to share files over a network. Windows file sharing and NFS are two of the more common file-sharing protocols.
- *FTP*: File transfer protocol (FTP) allows employees and other customers outside of the organization the ability to transfer files to and from a server. Often FTP services are embedded within web services so that files can be downloaded via a web browser.
- *Web services*: Provides the ability to serve up and out web pages. The World Wide Web (WWW) is based on the client–server model where clients (users) request a web page from a server.
- *Domain name service (DNS)*: Provides the ability to look up an IP address from a URL. When we use a web browser, we type in a URL such as http://www.apple.com. But the Internet needs to have an IP address (such as 136.15.82.4) in order to find the web page you are looking for. DNS provides this translation and does it automatically, expeditiously, and (usually) accurately.
- *Dynamic host control process (DHCP)*: Provides the ability to serve up IP address to hosts. All computers need a unique IP address in order to communicate via the Internet. These addresses can be manually assigned, which is very labor intensive and slow, or assigned by a DHCP server. When a computer turns on, it calls out to the network that it needs an IP address. The DHCP server hears the request and responds with an IP address for the computer to use.

Simple Network Management Protocol

In order for machines to communicate with other machines concerning their health, well-being, and management, a protocol is required. The most common protocol used for management of networks is the simple network management

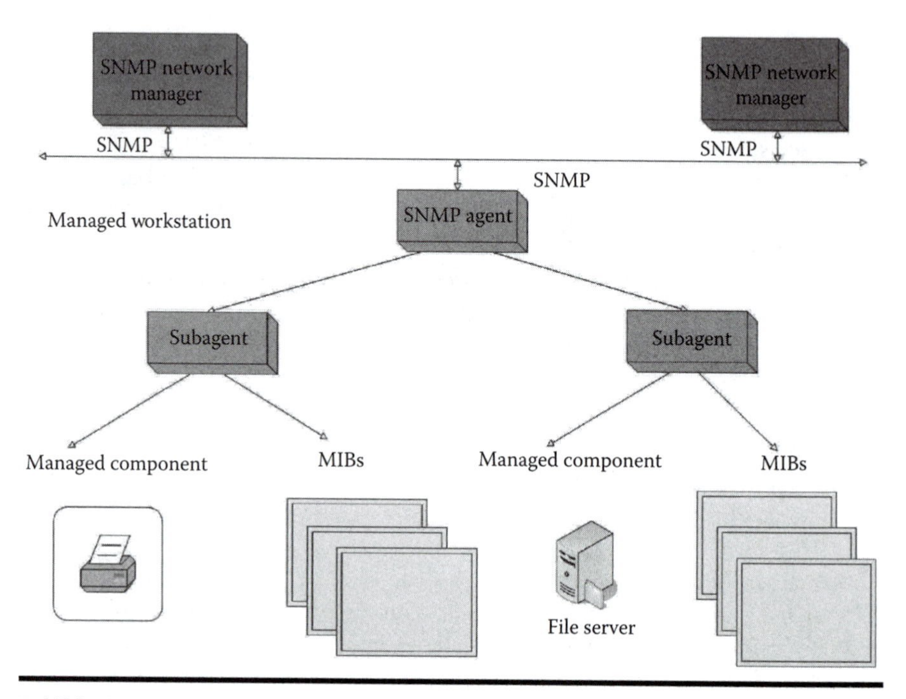

Exhibit 16.11 SNMP structure.

protocol (SNMP). SNMP allows for interoperability of management access among many different computer and device platforms.

SNMP is actually a set of standards. The standard includes a protocol, a database structure, and a set of data objects. SNMP was adopted as the standard for managing the Internet in 1989 and, with numerous upgrades, still serves as the standard today.

The organizational model for SNMP consists of four parts: the management station, management agents, management information base (MIB), and protocol. Exhibit 16.11 displays the SNMP structure.

Management Station

Usually a stand-alone computer that has management software riding on it. This management is the collection point for all information and makes requests of agents for information.

Management Agents

These are the innumerable devices on a network (computers, iPads, smart phones, routers, switches, etc.) that speak with the management station. An agent keeps track of and sends to the management station data such as state of network operation, number of error messages received, number of bytes and packets into and out of the device, broadcast messages sent, general health (temperature, fan operation, process speed), and interface flapping.

Management Information Base

The management station and agents store requested information in a MIB. The MIB is a structure (software design) that maintains information about a device within an agent (see Exhibit 16.12). For example, MIB slot 1843 is where information on the temperature of the device is stored. The management station can request MIB 1843, and the agent returns the value (temperature) to the manager.

Network Management Protocol

SNMP is a protocol and a suite of tools used to manage a network. As a protocol, there are set rules, frame structures, and commands that get used. The four

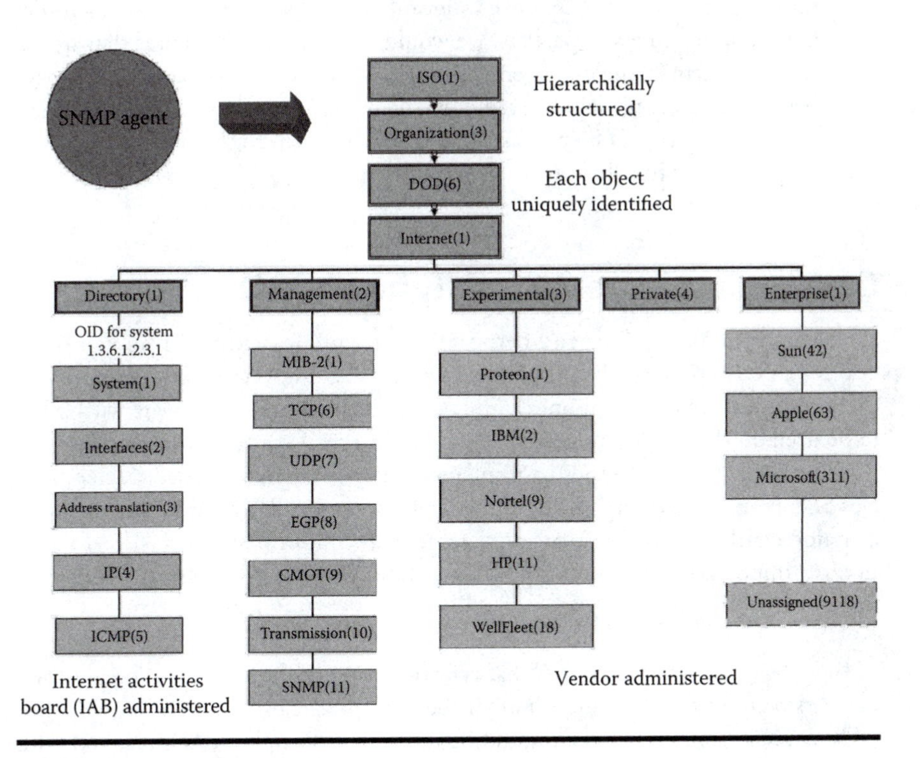

Exhibit 16.12 MIB structure.

commands that are common for SNMP are GET, GET NEXT, SET, and TRAP. These commands typically come from the management station to an agent.

- GET is a command that draws a data element from the MIB database. For example, GET 1843 draws that data element from the MIB. Data element 1843 happens to be the *operating temperature* of the device. So, a GET 1843 would obtain the operating temperature of the device called for.
- GET NEXT is a command that gets the next sequential data element in the MIB. So, using the example above, a GET NEXT command would get data element 1844 from the MIB.
- SET is a command that sets the MIB element back to a predefined value. For example, a SET 1990 "0" would take the value of MIB 1990 and set its value to a "0." Suppose that MIB value 1990 kept track of *packets received* on a particular interface; the SET 1990 command would return the value to "0" and therefore allow the management station to count from "0" from that point onward. This command allows manipulation of the MIB database that exists on the agents.
- TRAP is a command that allows for setting of predetermined value alarms on an agent. For example, if we want to know when the interfaces of a router are dropping excessive packets, we would set a trap on the router that would send an alarm to the management station if the number of packets that have been dropped exceeded 20 per hour. The management station tells the agent what to keep track of, and the agent sends an alarm to the management station when the threshold is reached.

Case Study: Network Security and BYOD

Increasingly, organizations and corporations are implementing bring your own device (BYOD) policies. BYOD is the concept of employees bringing and using their own devices (smart phones, tablets, portable computers) to the workplace to be used on the corporate network, often in place of corporate-owned devices. Such policies can be cheaper to implement than corporate-owned device policies and are more convenient for employees. However, BYOD policies represent a major challenge to security management and administration. The following are important aspects of a BYOD policy that must be well planned and managed:

- *Device choice and support*: What type of devices will the corporation allow on its premises? Will it support all of them (impractical)?
- *Bringing on new devices*: Should be done by the employee, simple, and with little ICT support.

- *Maintaining security to the corporate network*: The organization must establish and enforce the minimum security baseline that any device must meet to be used on the corporate network (Wi-Fi security, VPN access, antivirus program, etc.).
- *Company usage policies*: Access to the Internet, peer-to-peer file sharing, and applications use must be subject to different policies.
- *Pulling back access*: When a device is stolen or lost or an employee leaves the organization, how does ICT revoke access?
- *New attacks*: With the devices representing a larger number of *doors* into the network, how does ICT protect corporate assets?
- *Managing expectations*: The employees will expect the same level of quality, uptime, and reliability on the BYODs as they will from the wired networks. This is difficult to do.
- *Address exhaustion*: Will the company have enough IP addresses for each device?
- *Overlap in function*: Where does work end and pleasure begin? Will this be a gain in productivity or a loss?

These are many of the key issues with the BYOD movement, but not all. This is a serious challenge for the ICT departments, in administration and security, but one that must be met as more employees come to expect BYOD environments in the workplace.

Summary

In this chapter, we delved into the worlds of network security and network management. These are fields that are of key importance in today's IT environment and have key relevance for our personal lives as well.

We first looked at profiles of the perpetrators. Some do it just for kicks, others to increase their bank accounts. Whatever the reason, the rise of computer data theft is here, and it is prudent to watch all our bases. We next looked at the types of attacks that can occur.

The basics of security revolve around authentication, accounting, and encryption. These three elements provide the basis for the tools, tricks, and procedures in the security field. Authentication involves ensuring who you are, accounting involves watching what you did, and encryption involves hiding the data that we send and store.

We discussed some of the best practices for network security. These involved network design, use of firewalls, IDSs and IPSs, and honeypots. We examined host security and locking down the host computer. This involves OS security, software security, security policies, and antivirus programs.

We closed the chapter with a discussion of network management and the primary tools of client/server architectures and of SNMP.

Review Questions

True/False

1. A black hat is a person who attacks networks and PCs with a desire to steal, corrupt, and do harm to the data and the system.
   a. True
   b. False
2. Network accounting is very helpful in finding hackers.
   a. True
   b. False
3. Symmetric key encryption contains a public and private key.
   a. True
   b. False
4. If access is gained to the wiring or open port, a thief could sniff out password or other critical information from the network.
   a. True
   b. False
5. Firewalls come only as software programs.
   a. True
   b. False
6. A diskless client has no hard drives and relies on servers.
   a. True
   b. False
7. The GET command in a MIB database draws a data element.
   a. True
   b. False

Multiple Choice

1. Software that sits on the Internet analyzing web traffic is referred to as a
   a. Worm
   b. Cracker
   c. Cookie
   d. Sniffer
2. A person, by sending innumerable requests to e-mail servers, thereby locking e-mails to recipients, is committing
   a. A malware attack
   b. A social engineering attack
   c. A hacker attack
   d. A denial of service attack

3. Which one of the following is not classified as a biometric?
   a. Digital password
   b. Sound of your voice
   c. Fingerprint
   d. Blood vessel in the retina of your eye
4. With spoofing
   a. Sensitive information about sales figures and SSN can get stolen off the wire.
   b. One can change information that is being transmitted.
   c. The network becomes open to hackers.
   d. Your username and password can be located or seen.
5. Biometric scanning is used for
   a. Accounting
   b. Authentication
   c. Encryption
   d. Security
6. Asymmetric key cryptography
   a. Is a form of private key encryption
   b. Is the basis for all encryption algorithms
   c. Uses two keys while sending and receiving
   d. Branches out into public key encryption, public key signature, and key agreements
7. A honey pot is a device that
   a. Looks old and runs on a vulnerable operating system with easy access for break-in attacks
   b. Quarantines the packet
   c. Blocks worms
   d. Acts as a firewall
8. Which of the following is an element of host security?
   a. Operating system security
   b. Software security
   c. Spam/pop-up and cookie blockers
   d. All of the above
9. A management agent in SNMP
   a. Checks on the performance of the working staff
   b. Checks for the mismanagement of resources
   c. Keeps track of the network users
   d. Monitors network devices like routers and switches

10. Which of the following is not a function of IDS recognized signature from a virus, or a DOS attack, it warns the system administrator by e-mail, pager, or phone call?
    a. IDS detects a recognized signature of a virus or DOS attack.
    b. IDS warns the system administrator by e-mail, pager, or phone call.
    c. IDS prevents the virus, malware, or attack from spreading through the network.
    d. IDS monitors the network.

Short Essay Questions

1. List the types of computer attacks.
2. What is the role of private and public keys in encryption?
3. List the layers implemented in a depth-in-design network design.
4. What are the four elements of host security?
5. What are the key aspects of network administration and management?

Appendix: Answer Key to Chapter Questions

Chapter 1 Systems and Models of Communications Technologies: Shannon–Weaver, von Neumann, and the Open System Interconnection Model

Answer Key

Multiple Choice

1. B
2. D
3. D
4. D
5. D
6. D
7. B
8. C
9. C
10. B

Matching Questions

1. G
2. A
3. D
4. I
5. C
6. H
7. E

8. B
9. J
10. F

Short Essay Questions

1. According to the Shannon-Weaver model of communication, a source sends an encoded message to a receiver that decodes the message, which passes through a channel that is susceptible to noise. In terms of human communication, a person may speak English to another person who understands English, but the speaker's words may be muffled by the wind.
2. Using a standard model in which to discuss networks or computers allows for many different vendors to create products that are compatible. If no standards existed, it would be extremely difficult to create networks or computing devices that are compatible because each vendor would have its own way of doing things, and that way may not be compatible with another vendor's method.
3. The components of the von Neumann stored-program concept are as follows:
   a. Main memory: the working area of the computer; stores data and instructions.
   b. Arithmetic logic unit: performs computational functions on binary data.
   c. Control unit: interprets the instructions in the memory and executes.
   d. Input/output equipment: devices operated by the control unit, such as a printer.
   e. Bus structure: the interconnection medium that allows all communications to occur within a computer.
4. The layers of the OSI model are as follows:
   a. Layer 7 (application): Provides a final integrity check for the data (user interface, FTP, HTTP).
   b. Layer 6 (presentation): The data translator and interpreter (form, syntax, encryption, and compression).
   c. Layer 5 (session): Establishes, manages, and terminates connections between devices (resource control).
   d. Layer 4 (transport): Responsible for end-to-end integrity of communications (TCP).
   e. Layer 3 (network): Logical addressing (IP) also formats data from higher layers into packets that are passed to the data link layer.
   f. Layer 2 (data link): Concerned with physical addressing (MAC).
   g. Layer 1 (physical): Describes how data are sent through a network (mechanical, electrical, bits).
5. The von Neumann model is the foundation of all digital devices. Originally presented in 1946, it can be used to explain how digital devices work. If every device had a different method of storing, collecting, modifying, or

transmitting information, designing networks and computing devices would be extremely difficult. The von Neumann model allows every computing device to be analyzed in terms of the components explained in the model.

6. The steps of the communication process are as follows:
   a. A source: a man speaking over the radio.
   b. Transmits a message: his words.
   c. That is encoded: the language he is speaking.
   d. Through a channel: the medium he uses (i.e., telephone, air, or radio waves).
   e. That is susceptible to noise: anything that changes or alters the original message; in the case of radio, atmospheric conditions can weaken a signal.
   f. The message is decoded: by the receiver; a person listening to the radio understands the language of the source.
   g. By the receiver: a person listening to the radio.
7. The fifth component of the stored-program concept is the bus structure. It is not often included in many diagrams explaining the concept, but it is important. The bus structure is critical for any computing device because it connects everything. It is the physical path upon which signals are carried.
8. The theory of Claude Shannon explains the relationship between the signal and noise in a given transmission based on the bandwidth or frequency at which the signal is being transmitted. The capacity of information being transmitted is determined by all of these factors. The formula demonstrates that as the noise in a transmission increases, the capacity to send information decreases, and as the frequency (or bandwidth) in which we transmit increases, we have a greater capacity for information transfer.
9. Answers will vary.
10. As a professional in the IT field, one must be able to speak the language. Even if you are not designing or developing new technologies, you are managing people who do. To effectively lead requires more than management experience; one must also have an understanding of the basic concepts that comprise the IT world.

Chapter 2 Basic Concepts of Electricity

Answer Key

Multiple Choice

1. D
2. E
3. A
4. C
5. D
6. D
7. B

8. A
9. E
10. D

Matching Questions

1. 1B	2D	3C	4A
2. 1C	2A	3B	4D
3. 1D	2B	3A	4C
4. 1C	2A	3D	4B
5. 1B	2D	3A	4C
6. 1A	2C	3D	4B
7. 1A	2D	3B	4C
8. 1D	2B	3A	4C
9. 1D	2C	3B	4A
10. 1B	2A	3D	4C

Short Essay Questions

1. A waveform is a picture produced by a specific piece of measuring equipment (e.g., an oscilloscope) that shows the highs and lows of the voltage or current associated with the signal. The waveforms vary in intensity, shape, duration, and complexity.
2. The frequency (unit: Hertz, Hz) of a signal is defined as the number of cycles divided by the time in which they occur. The period (unit: second, s) is the time it takes a waveform to complete one complete cycle.
3. Bandwidth is a range of frequencies a communication channel is defined within for specific signaling functions. As an example, telephone company circuits operate over a range of frequencies from 300 to 3400 Hz.
4. Current flow (unit: ampere, A) is the movement in the same direction in a conductor of the free elections. Alternating current (AC) is the type of electrical power we have at our wall outlets. The other type of current flow is direct current (DC). AC can be thought of as changing direction as it goes from its positive position to its negative one; DC can be considered a continuous flow of current as opposed to the periodic condition of AC.
5. Resistance (unit: ohm, Ω) is the force that opposes the flow of electrons in a material.
6. The ability to do work or the energy potential of the electrical charges is called voltage (unit: volt, V).
7. The ability of an electrical conductor to hold a charge is considered its capacitance.
8. Inductance is the resistance of a conductor to allow current to change direction.

9. When electrons are forced to move between points of potential difference (e.g., positive and negative terminals on a battery), work is being accomplished. The measure of the rate at which work can be accomplished is called power (unit: watt, W).
10. DC is derived from AC by a device known as a rectifier, a device that uses special components to control the flow of AC current and bring the negative or alternating side of the signal together with the positive side of the signal. Devices called diodes allow current to flow through them in one direction only. The diode acts like a switch to current. Coupled with a capacitor, diodes can create a constant current flow for components within electronic equipment.

Chapter 3 Modulation Schemes

Answer Key

True/False

1. False
2. False
3. True
4. True
5. False

Multiple Choice

1. A
2. C
3. C
4. A
5. C
6. A
7. B
8. B
9. A
10. B

Short Essay Questions

1. The Federal Communications Commission.
2. Government (navigation, military, aviation, secure communications, and public safety); licensed (cellular, wireless local loop, satellites, radio and television stations); and unlicensed (ISM, cordless phones, wireless Internet, and microwave ovens).

3. It can be multiplexed or modulated.
4. AM radio signals use a longer wavelength and a lower frequency; this, combined with the ionosphere acting as a reflectant, allows the signal to travel much further than FM.
5. Because power or amplitude is modulated if the signal encounters a power spike, that spike is added to the signal in the form of static and interference.
6. The higher the frequency, the shorter the wavelength; the lower the frequency, the longer the wavelength.
7. Amplitude modulation uses the waveform to modulate the amplitude or power. Frequency modulation uses amplitude as the carrier for the frequency.
8. Amplitude-shift keying (ASK) and phase-shift keying (PSK).
9. An electromagnetic signal that is modulated in amplitude, frequency, or phase to transmit information.
10. Adaptive differential pulse-code modulation (ADPCM).

Chapter 4 Signaling Formats, Multiplexing, and Digital Transmissions

Answer Key

Multiple Choice

1. B
2. A
3. B
4. B
5. A
6. D
7. C
8. C
9. A
10. C

Matching Questions

1. D
2. J
3. H
4. C
5. I
6. E
7. G

8. A
9. B
10. F

Short Essay Questions

1. Half duplex allows for transmission in one direction at a time, whereas full duplex sends information in both directions at the same time. If in a cost-prohibitive situation, half duplex might be more economical. Full duplex would be a more logical choice for high transmission. An example of full duplex is a phone conversation or a videoconference.
2. An advantage is the consistency by which the circuit checks for errors; a disadvantage is the ease with which an error can occur. In AMI, you have an effective technique to detect errors. The prior four binary formats will detect an error only if there is a consecutive string of 0s.
3. Bipolar RZ (AMI) is the most reliable signaling format. This process is used because of the power requirements necessary for transmission and the ability to detect errors in the signal stream more effectively. When a bipolar violation occurs, the circuit is immediately made aware without the use of any more energy.
4. ZCS and B8ZS are keep-alive signals. In ZCS, the signal is kept alive by inserting a mark in a string of eight consecutive 0s, which causes a bipolar violation that keeps the signal alive. In B8ZS, the signal is kept alive by inserting a code into the string of 0s. This is done by placing a mark at the fourth, fifth, seventh, and eighth intervals.
5. The difference between the three formats is that AMI is used for error detection, while ZCS and B8ZS are used as keep-alive signals.
6. The CSU/DSU must be programmed to the particular signaling format that you are looking to use. If a digital circuit is ordered with ZCS keep-alive signaling while the carrier provides B8ZS, alarms, faulty operation, and flaky problems start appearing in relation to the circuit. If your system is looking for ZCS and a packet of 1s and 0s comes down the line as B8ZS, the system tries to analyze the data bits as information.
7. If a digital circuit is ordered with ZCS keep-alive signaling while the carrier provides B8ZS, alarms, faulty operation, and flaky problems start happening in relation to the circuit. If the system is looking for ZCS and a packet of 1s and 0s comes down the line as B8ZS, the system tries to analyze the data bits of information.
8. Both FMD and TMD are techniques used to send information in a serial transmission. The way in which they multiplex is different; FDM separates the information by frequency, while TDM is a more widely used method because the synchronization of the TDM signal is critical in the delivery of information. Each time slot requires special framing and coordination bits

that tell the receiving end where each information slot stops and starts. TDM is based on digital signaling from end to end.

9. A major disadvantage of TDM is the waste of available time slots. When a particular channel has no information to send, that time slot will contain nothing more than a zero. This wasted space makes this transmission method inefficient in some respects.
10. 24 DS-0s × 64,000 kbps (the throughput of 1 DS-0) = 1.536 Mbps. You then add back the 8000 bits used for sampling for a total of 1.544 Mbps.

Chapter 5 Legacy to Current-Day Telephone Networks

Answer Key

True/False

1. False
2. True
3. False
4. True

Multiple Choice

1. A
2. D
3. D
4. B
5. C

Short Essay Questions

1. Public switched telephone network.
2. Voice over Internet Protocol.
3. SS7, a frequency band, different from the path taken by the transmitted signal, which contains the circuit acquisition, setup, and teardown information for that signal.
4. ATM is a compromise between circuit switching and packet switching. It is a cell-based, asynchronous delivery technology designed to carry voice, data, and multimedia services over the same link simultaneously. It can emulate a traditional voice network and provide asynchronous data transfer over the wide area.
5. The telephone, network access, central offices, trunk and lines, and customer premise equipment (CPE).

Chapter 6 Basics of Multiprotocol Label Switching Networking

Answer Key

True/False

1. False
2. True
3. True
4. True
5. True

Multiple Choice

1. D
2. A
3. A
4. D
5. B

Matching Questions

1. G
2. B
3. A
4. C
5. H
6. E
7. D
8. F

Short Essay Questions

1. Multiprotocol Label Switching (MPLS) is a switching approach meant to implement ATM-like virtual-circuit-equivalent approaches into an IP overlay on existing networks.
2. It is similar to ATM's call setup message submitted by an ATM network user, which requests that a certain quality of transmission service be set up across an ATM network. They have a similar packet forwarding and path reservation. Both disseminate path information in a variation of IP's open-shortest-path route-dissemination procedure.
3. Because the MPLS label is inserted between the layer 2 and layer 3 addresses.

Chapter 7 Local Area Network Technology

Answer Key

True/False

1. False
2. True
3. False
4. False
5. True
6. True
7. False

Multiple Choice

1. C
2. B
3. A
4. A
5. B
6. A
7. D

Short Essay Questions

1. CSMA/CD stands for Carrier Sense Multiple Access/Collision Detection. It is used with Ethernet networks and works as a contention-based protocol. The computers on the network listen until they hear nothing (meaning the bus is clear), and then they put their packets on the bus (the CSMA part). If the bus is busy with another computer's data, the computer waits until the bus is clear before it transmits. A collision occurs when two computers put their packets on the bus at the same time. When a collision occurs, all computers stop all transmissions and then wait for a random time period before starting transmissions again.
2. The four components of a LAN are (1) the end-user devices, (2) the physical media, (3) the networking equipment, and (4) the network operating system (NOS). The end-user devices are network-connected devices that the users see and interface with. They are the transmit-and-receive ends of the human side of the network. Examples of shared end-user devices are printers, fax machines, and scanners. The physical media is the wiring between the end-user devices that allows the communication to occur. The type of media used depends on what is currently existent in the building or the characteristics of the organization. The networking equipment is what allows the network

to send information to the correct destination. It is basically the core of the network. Switches, routers, and NICs are the most common types of network equipment. The NOS controls the order of communication between end-user devices (both personal and shared). Unlike a PC operating system that controls an individual piece of equipment, the NOS controls communication between the devices.

3. Packet switching is an efficient use of resources where the path (or circuit) is shared by many users at once. Information is sent in packets, and each packet carries information that is destined for different locations. In circuit-switched networks, a dedicated path is carved out of a larger network to form a communication path. This path or circuit is set up by signaling created at the time you want to communicate and terminated at the time the communication is ended. The entire circuit is reserved during the length of the call.
4. Two broad categories of network costs are initial and ongoing costs. Initial costs are incurred during setup and installation, including costs for wiring, equipment, network devices, and software. Ongoing costs are those incurred during the life of the network, including maintenance, personnel, MACs, and licensing.

Chapter 8 The Language of the Internet: Transmission Control Protocol/Internet Protocol (TCP/IP)

Answer Key

True/False

1. True
2. False
3. False
4. True
5. False
6. True
7. True
8. False
9. False
10. True

Multiple Choice

1. B
2. A
3. B

4. C
5. C
6. B
7. C
8. C
9. B
10. D

Short Essay Questions

1. The primary advantage of centralized routing is network efficiency. Centralized routing uses one routing table that keeps track of the whole network. As such, computer resources and network capacity are used by one managing device only.
2. The session is ended when one of the logical units (end-user device) sends a deactivation request.
3. The languages of the network, routed protocols, operate at the network layer of the OSI model. They form the information that is communicated between computers.
4. TCP/IP is the most common routed protocol. It is made up of two separate parts: TCP and IP. The primary function of the IP part is to address and route the packets. The other part, TCP, has the function of packing and unpacking the data (reassembling the data packets) and ensuring reliability.
5. Multicast messages are transmitted by sending a message to people in a distinct group or classification. Only those to whom the message is addressed can hear the broadcast message.

Chapter 9 Wireless Local Area Networks

Answer Key

True/False

1. False
2. True
3. True
4. False
5. False
6. True
7. False
8. True
9. True
10. False

Multiple Choice

1. E
2. D
3. C
4. A
5. D
6. B
7. A
8. C
9. D
10. C

Short Essay Questions

1. Wireless LANs provide physical flexibility and mobility. If a company needs to physically move people and computers, wireless LANs provide instant mobility. If a company is housed in a building with a structure that is difficult or dangerous to change, or if the building is protected by historical preservation laws, a wireless LAN can take the place of drilling through walls.
2. Ad hoc connected computers talk only to each other, whereas computers that are functioning in infrastructure mode connect to a central access point in a logical star topology. In infrastructure mode, computers can have access to the local LAN and possibly to the outside Internet; in ad hoc mode, this is not possible.
3. Omnidirectional antennas are advantageous when connectivity must come from all directions. If the connecting computers are on all sides of the access point, an omnidirectional antenna can connect with everyone. If the access point is in a corner, a unidirectional antenna can focus the connections toward the connecting computers.
4. Interference can be caused by walls, buildings, people, objects, or anything that the radio signal must travel through or around. In multistoried buildings, interference can be caused by ceilings and floors. Radio interference can be caused by anything that transmits electromagnetic waves, including electric devices (motors), area radio towers, and even mobile phones.
5. Wireless LANs are easy to break into if no security precautions are set up. Anyone with a wireless card can hop onto a wireless network unless protections such as WEP are in place. These security issues include access to sensitive material as well as use of purchased uplinks.

Chapter 10 Mobile Wireless Technologies

Answer Key

Multiple Choice

1. D
2. C
3. C
4. B
5. C
6. A
7. D
8. A, C
9. A, D
10. B

Matching Questions

1. 1B	2D	3C	4A
2. 1C	2D	3A	4B
3. 1C	2A	3D	4B
4. 1B	2A	3D	4C
5. 1A	2C	3D	4B
6. 1D	2C	3B	4A
7. 1C	2A	3D	4B
8. 1B	2C	3D	4A
9. 1D	2B	3A	4C
10. 1A	2C	3D	4B
11. 1B	2C	3A	4D

Short Essay Questions

1. From its inception, AMPS has been referred to as a cellular service because of the configuration of the antenna propagation field. Theoretically shaped like honeycomb cells on the engineering layout, the cell sites provide a way to reduce power, increase access, and reuse frequencies in the limited bandwidth allotments for cellular services. The reuse-of-seven rule, which states that no cell can use a frequency from an adjacent cell, forced diversity in frequency placement in the network.
2. Frequency division multiple access (FDMA) is a multiple-access technique in which users are allocated specific frequency bands. The user has a singular right of using the frequency band for the entire call period.

3. Time division multiple access (TDMA) is an assigned frequency band shared among a few users. However, each user is allowed to transmit in predetermined time slots.
4. Code division multiple access (CDMA) is a technique in which users engage the same time and frequency segment and are channelized by unique assigned codes.
5. The mobile switching center (MSC), which is known by a number of different names, including mobile telephone switching office (MTSO), provides the system control for the mobile base stations (MBS) and connections back to the PSTN. The MSC is in constant communication with the MBS to coordinate handoffs (or switching) between cell sites and connections to the landlines, and to provide database information about home and visiting users on the network.
6. An antenna is a circuit element that provides a changeover from a signal on a transmission line to a radio wave and for the gathering of electromagnetic energy (i.e., incoming signals). An antenna is a passive device in the network because it is a receiver and transmitter of electromagnetic energy but is not responsible for amplifying the signal. An antenna is defined by four characteristics: reciprocity, polarization, radiation field, and gain.
7. Smart antennas are base station antennas with a pattern that is not fixed but adapts to the current radio conditions. Smart antennas include the switched lobe (also called switched beam), dynamically phased array, and adaptive array.
8. Smart antennas add a new way of separating users, namely, by space, through space-division multiple access (SDMA). This can be visualized as the antenna directing a beam toward the communication link only. SDMA implies that more than one user can be allocated to the same physical communications channel simultaneously in the same cell, only separated by angle.
9. Microwave signals are radio-frequency (RF) signals that range from 1 to 40 GHz. As we move up the spectrum, the radio waves become susceptible to conditions that affect the transmission of light. Frequencies below 6 GHz are not greatly affected by line-of-sight issues or obstructions blocking signals. Beyond 6 GHz, we are constrained by issues of distance forced on us by the curvature of the earth, atmospheric conditions such as fog and rain, and buildings and other structures that could impede the signal's transmission path.
10. The bit error rate (BER) is a performance measure of microwave signaling throughput. It represents the number of pieces of information corrupted or lost during transmission. There are four primary diversity issues that need to be considered with microwave placement: space diversity, frequency diversity, hot standby, and the use of a PRI connection for a failover. The mean time between failures (MTBF) of the equipment being used is an estimated time frame in hours for the durability of the equipment, provided by the manufacturer.

Chapter 11 Virtualization

Answer Key

True/False

1. True
2. False
3. True
4. True
5. True

Multiple Choice

1. B
2. C
3. A
4. D
5. A

Short Essay Questions

1. In RDS, users are connected to a virtual machine hosted on a remote VDC server. All RDS users remote into the same VM operating system. In VDI, each user is allocated his or her own virtual machine hosted on a remote VDC server. VDI users remote into separate operating systems.
2. Thin LUNs are created from a thin pool of physical storage drives and allocated to virtual machines.
3. Storage virtualization at the file level provides abstraction in the process of separating files from their actual physical locations on physical disks. File-level virtualization simplifies file mobility by creating a logical pool of storage enabling users to employ a logical path in place of a physical path to access files.

Chapter 12 Analyzing Big Data

Answer Key

True/False

1. True
2. True
3. False
4. False
5. True

Multiple Choice

1. B
2. C
3. C
4. B
5. D

Chapter 13 The Cloud and Cloud Computing

Answer Key

True/False

1. True
2. False
3. True
4. False
5. True

Multiple Choice

1. C
2. A
3. C
4. D
5. E

Short Answer

1. A style of computing in which scalable and elastic IT-enabled capabilities are delivered as a service using Internet technologies.
2. On-demand self-service, shared resources, rapid elasticity, and measured service.
3. The hybrid approach allows a business to take advantage of the scalability and cost-effectiveness that a public cloud computing environment offers without exposing mission-critical applications and data to third-party vulnerabilities.

Chapter 14 Video Basics

Answer Key

True/False

1. True
2. False
3. False
4. True
5. False
6. True
7. True
8. True
9. False
10. True

Multiple Choice

1. C
2. B
3. A
4. D
5. A
6. C
7. A
8. B
9. D

Short Essay Questions

1. It is easier to stop or inhibit noise (interference) in the system, allowing a cleaner signal that is free of jitter and other elements we are accustomed to seeing in a normal analog TV. Enhanced image: In a digital format, one can manipulate the images easily, allowing new capabilities, e.g., picture in picture, carrying 5.1 audio.
2. Compression is a process by which a mathematical computation is placed on a video and audio file. This process decreases the bandwidth necessary to carry the digital television transmission and enables the information to be carried in the original 6 MHz bandwidth. The digital broadcasting utilizes the MPEG-2 compression format in its transmission.
3. The three basic elements of video conferencing are the sender, its transmission media, and the receiving station.

4. With film pictures (as seen at the movie theater), the complete picture is projected in front of us at a moving rate of 45 images per second; video pictures (in the form of TV images) are scanned, line by line, at a high rate of speed (6 million bits per second) to give a visual image rate of 60 images per second.
5. In order for the carrier to contain the complete television signal, it must carry the following:
   a. Luminance information: picture brightness
   b. Synchronization pulses: synchronize the transmitter and receiver
   c. Blanking pulses: tell the image when to turn on and off
   d. Audio: sound information
6. The three primary additive colors are red, blue, and green. The three subtractive colors are the following:
   a. Cyan: a combination of blue and green
   b. Yellow: a combination of red and green
   c. Magenta: a combination of blue and red
7. The three essential qualities of videoconferencing systems are the following:
   a. High quality: The transmission system is called full frame and achieves the quality expected from professional studios. Transmissions are sent over high-bandwidth media and are typically fiber optics or satellite. Utilizes high-resolution TV and high-quality speakers on the receiving end. High cost.
   b. Medium quality: Compromises the video and sound quality in order to carry the image over limited bandwidth. Video quality is less than commercial TV quality. Medium-priced VC systems, with lower-priced cameras and microphones. Video and audio are compressed to enable travel over the bandwidth.
   c. Low quality: Images are jerky, and audio is harsh. Equipment is low cost and available for use in homes and home offices. Typically used over a broadband connection such as cable modem and DSL. Low cost and low quality.
8. Persistence of vision is the rapid presentation of frames of video information that give the illusion of smooth motion.
9. Image scanning consists of *looking* in horizontal passes across the target image and measuring the instantaneous amplitude that represents the reflected light energy. Each pass across the image by the scanning element is called a scan line. The video camera contains an image tube, or target, that automatically scans the scene and generates a corresponding electric output that replicates, in electronic form, the image that the camera sees.

Chapter 15 Digital Media

Answer Key

True/False

1. True
2. False
3. False
4. False
5. True
6. True
7. False
8. True
9. False
10. False

Multiple Choice

1. B
2. A
3. C
4. D
5. A
6. D
7. A
8. A
9. B
10. D

Short Answer Questions

1. Digital photography, digital video, digital audio, digital gaming.
2. Color filters are used to capture the different sensitivities of either the additive or subtractive primary colors. If the camera is additive, the additive primary filters red, green, and blue are used. RGB added together makes white light. For the subtractive concept, it is cyan, magenta, and yellow filters that are used. Cyan, magenta, and yellow together make black.
3. Speed.
4. Dye sublimation, ink jet, and laser.
5. The head end, satellites, and receiver.

Chapter 16 Network Security and Management

Answer Key

True/False

1. True
2. True
3. False
4. True
5. False
6. True
7. True

Multiple Choice

1. D
2. D
3. A
4. A
5. B
6. C
7. A
8. D
9. D
10. C

Short Essay Questions

1. Social engineering, denial of service, malware, and sniffing.
2. Data are encrypted by the public key and can only be decrypted by the private key. In asymmetric key encryption, the private key is kept secret, while the public key may be widely distributed.
3. External, DMZ, protected zone 1, and protected zone 2.
4. Operating system security, software security, virus protection, spam/pop-up and cookie blockers.
5. ICT auditing and assurance, backup and disaster recovery planning, service-level agreement (SLA), sustainability, and management of personnel.

Index

A

D

N

O

P

Q

T

W

X